Broadband Return Systems for Hybrid Fiber/Coax Cable TV Networks

Donald Raskin
Dean Stoneback
NextLevel Systems

ISBN 0-13-636515-9

Prentice Hall PTR
Upper Saddle River, NJ 07458
http://www.phptr.com

Library of Congress Cataloging-in-Publication Data

Raskin, Donald.
Broadband return systems for hybrid fiber/coax cable TV networks/
Donald Raskin, Dean Stoneback.
p. cm,
Includes index.
ISBN 0-13-636515-9
1. Cable television. 2. Broadband communication systems.
3. Coaxial cables. 4. Fiber optic cables. I. Stoneback, Dean.
II. Title.
TK6675.R37 1997 97-47131
621.388'57--dc21 CIP

Editorial/production supervision: *Dit Mosco*
Cover design: *Scott Weiss*
Cover design director: *Jerry Votta*
Manufacturing manager: *Alan Fischer*
Marketing manager: *Betsy Carey*
Acquisitions editor: *Bernard M. Goodwin*

Prentice-Hall, Inc.
A Simon & Schuster Company
Upper Saddle River, New Jersey 07458

Printed in the United States of America
10 9 8 7 6 5 4 3 2 1

ISBN 0-13-636515-9

Prentice-Hall International (UK) Limited, *London*
Prentice-Hall of Australia Pty. Limited, *Sydney*
Prentice-Hall Canada Inc., *Toronto*
Prentice-Hall Hispanoamericana, S.A., *Mexico*
Prentice-Hall of India Private Limited, *New Delhi*
Prentice-Hall of Japan, Inc., *Tokyo*
Simon & Schuster Asia Pte. Ltd., *Singapore*
Editora Prentice-Hall do Brasil, Ltda., *Rio de Janeiro*

To my wife, Mary

Once you have found her, never let her go

And to Carolyn and David

Lady Elaine

D.R.

To my wife, Pam

I praise God for bringing us together.
You are an unbelievable gift and
I will cherish you forever.

D.S.

Contents

Preface

Television engineers, designers, and operators are working to make the return path system—the conduit for signals from subscribers' homes toward the headend—as effective and efficient as the forward path already is. The forward path in today's hybrid fiber/coax (HFC) cable architectures routinely provides 60 to 110 analog channels of entertainment video. In addition, these cable TV systems can now provide individual households with services that are customer-specific, such as Internet access, interactive games, and near-video-on-demand, without incurring the high initial investment cost required for switched delivery systems. Although local telephone lines can be used for upstream requests from cable subscribers for information and programming, the cable TV return system provides the pathway to fully interactive communications services that are user-friendly, high-speed, and cost-efficient.

The use of the return path is in its infancy, however, even though two-way cable TV technology has been available since the early 1970s. The set-up procedures and performance specifications that are well established for the downstream system are not yet as clearly determined for the upstream. Also, as systems are being planned and constructed, there needs to be a full understanding of the equipment and service quality requirements so the new interactive services can be offered reliably, but without unnecessary cost. In addition, many electronic equipment designers, including engineers at companies unfamiliar with the cable TV industry, are pursuing the business opportunities offered by the new cable services—such as cable modems, HFC telephony equipment, and interactive terminals—and need to increase their knowledge base in this new field.

Until recently the design of cable TV systems was determined almost completely by the requirements of the forward transmission path. This is because—until recently—there was little or no revenue potential in return transmissions. That design bias is already changing. As an example, cable operators are designing and building new plant with fiberoptic service distribution areas sized for their return band capacity, which means that more nodes are being deployed than would be required for forward services only. The following chapters will establish some guidelines for determining capacity and will explain additional design considerations brought on by full two-way digital services.

This book attempts to answer the questions that are being raised by at least three different groups associated with the cable TV industry.

By cable operators:

- What services can be provided?
- What equipment is used to deliver each application?
- How much equipment will be required?
- How many subscribers for each service can be served from each fiberoptic node?
- How many subscribers can be served by one headend fiberoptic receiver?
- How many subscribers can be served by each application receiver?
- What is the configuration of a multi-service headend?
- What are the end-to-end performance requirements for the different services?
- What are the set-up and maintenance procedures for the return transmission system?

By system designers:

- What equipment is called for in the headend, in the distribution, and in the home?

- What are the choices in system architecture and how are they decided?
- How are the performance capabilities and design margins determined?
- How are network management systems configured?
- What are the useful precautions?

By equipment designers:

- What are the performance requirements for the individual pieces in the headend, in the distribution, and in the home?
- What are the principal characteristics of the transmission channel?
- How are the services managed?

Since the subject matter covered in this book is quite broad, our approach is to establish clearly the basis for the technical discussion in each case. We realize that this entails the risk of some readers finding parts of the discussion to be too basic, and we apologize in advance to those readers. We feel that it is important to avoid the opposite risk: not addressing the needs of all the potentially interested readers by failing to build a sufficiently broad foundation.

In the interests of keeping the discussions as straightforward as possible, we have had to make a small number of simplifying omissions. We discuss only return signals that are in a frequency band *below* the forward signals, thus ignoring the potentially useful idea of placing the return band above the forward, as was deployed in Time Warner's Full Service Network trial in Orlando, Florida. We have also assumed that only *digital* signals are carried back to the headend on the return path. It is clear that there are situations when an operator needs to transport one or two analog video signals upstream (for instance, to acquire city council sessions or educational TV programs for re-transmission over the downstream network), and it would seem logical to attempt to share the return band between the analog signals and the digital ones. In reality, however, this would mean foregoing the service revenue potential of the digital signals that could be carried in the 6 MHz required for each analog channel. Hence it will generally be much more cost-effective to run a separate dedicated fiber from the video origination point.

We have included four appendices. The first, for people who are used to analog signals, describes how to make measurements of digital signals with a spectrum analyzer. The second provides details on some of the set-up items treated in Chapter Eleven. That appendix also discusses how to transport analog video on the return path, for situations where this is required. The third appendix is a brief introduction to the baseband digital structures of the public switched network (SONET/SDH and ATM) and to the protocols of the Internet (TCP/IP and UDP/IP). The fourth contains tables and conversion formulas useful in cable TV calculations.

The Engineering Committee of the National Cable Television Association (NCTA) is supplementing the *NCTA Recommended Practices for Measurements on Cable Television Systems* to include issues and techniques related to upstream transport.[1] One of the authors (D.S.) has been heavily involved in drafting those recommendations. Since the NCTA document is a full and detailed description of testing procedures, we have not dealt with that subject in this book.

Last, it must be acknowledged that it has not been possible to include detailed discussions of all types of equipment for all types of two-way applications. In part this is to keep the book to a reasonable length; in part it is because both the applications and the associated equipment are evolving. The emphasis in the book is to establish the fundamentals underlying the applications and to develop a solid understanding of the return transmission network. It is our hope that this foundation will endure as the technology evolves and as the service applications flourish.

Donald Raskin
draskin@nlvl.com

Dean Stoneback
dstoneback@nlvl.com

1. The supplement can be downloaded from the CableLabs web site: `http://www.cablelabs.com`.

Acknowledgments

Among our colleagues at NextLevel Systems,[1] the working title for this book has been "Ken Simons Returns," in deference to the forward path text by that author, which our company has published since 1965. In addition to setting that high standard for us, our co-workers and our friends from other companies and cable operations have given us considerable support throughout this book's preparation.

We owe a deep debt of gratitude for the insights shared by our fellow employees in the Transmission Network Systems group of NextLevel's Broadband Networks Group, including Fred Slowik and Curt Smith in Systems Marketing; Jon Weinstock and Ron DiUbaldi in Network Management Development; Tim Funderburk, Frank McMahon, Rob Howald, and Mike Aviles in Systems Engineering; David Ciaffa, George Scherer, and Tim Homiller in Design Engineering; Charles Breverman, Rudy Menna, and Tim Brophy in Advanced Technology; and Julie Hansberry in Marketing. We benefited greatly from discussions on set-up and operation with Bill Beck, Chuck Moore, Walt Sharp, Helmut Hess, John Ridley, and Verne Johnson. Vipul Rathod was our lifeline to reality through

1. Some readers will be more familiar with the former company name, General Instrument Corporation, or the preceding name, Jerrold Communications.

his thorough lab work. Steve Vendetta supplied some informative drawings, and Jennifer Johnson helped translate drawing formats. We would also like to thank our management, Tom Lynch, David Grubb, and Chuck Dougherty, for their support and encouragement.

We received valuable information and valued insights from Henry Blauvelt of Ortel; from Stuart Lipoff of Arthur D. Little; from Randy Goehler of Cox Communications (San Diego, CA); from James Waschuk of Shaw Cablesystems (Calgary, Alberta); and from Tony Werner and Oleh Sniezko of TCI Engineering (Englewood, CO). Clive Holborow and Jim Green of NextLevel's SURFboard cable modem group helped us understand this key return path application and its transport requirements.

We have benefited from the active exchange of correspondence over the "SCTE-List," an e-mail forum for the discussion of cable television and telecommunications technology operated by David Devereaux-Weber of the University of Wisconsin–Madison's Division of Information Technology.[2]

We would like to express our sincere appreciation to Randy Goehler, David Devereaux-Weber, Ron Hranac (Coaxial International, Denver, CO), and Clive Holborow for their careful and critical reading of the draft manuscript and for their valuable suggestions.

Last, because the material in this book is based on the hybrid fiber/coax (HFC) cable TV architecture, we need to recognize the work of all of the people who made that network technology possible. Most particularly, we and the industry owe a great debt to Jim Chiddix, Dave Pangrac, Louis Williamson, and others at Time Warner Communications for pressing the development of semiconductor lasers suitable for analog transmission, and to Bill Lambert and George Fletcher for having conceived the use of multi-port nodes and amplifiers for multiple-star signal distribution.

2. The SCTE-List has no official connection with the Society of Cable Telecommunications Engineers. Interested parties may subscribe by sending the message
`subscribe scte-list firstname lastname`
to the host
`listserver@relay.doit.wisc.edu`
There is no charge for a subscription.

CHAPTER 1

Introduction

It can be argued that the future of the broadband cable industry in the United States hinges on how well the industry exploits the capability of its return path. The reasoning goes as follows:

- Penetration of basic and premium analog video services is unlikely to increase because the nation has been "wired" for several years, and competitive delivery systems that offer similar downstream services are emerging.
- Cable delivery of *digital* video to the home is a necessary service enhancement that will strengthen the cable industry's ability to compete against other offerings, such as MMDS[1] and DBS,[2] but its primary effect will be to retain market share, not to grow it.
- The key differentiator between cable and other delivery systems is the existence of a broadband *bi-directional* "pipe" between the headend and subscribers' homes.
- A number of service applications require such a broadband bi-directional pipe, for which strong market demand has been demonstrated.

1. Multichannel Multipoint Distribution Service, a terrestrial microwave distribution.

2. Direct Broadcast Satellite, distribution via microwave satellite relay.

The authors agree with this line of reasoning. We know that others hold different views. In fact, it has become routine to hear pundits declare—whenever one or another high-profile trial is abandoned by either a cable operator or a telephone company—that interactive broadband communications is a dead issue. In our opinion, these experts are not reading the market correctly, for the following reasons.

- Even though we are just beginning to grasp the power of the Internet, it is clear that users—a broad range of users—need better access through higher-speed file transfer. This requires bi-directional broadband transport.
- Anyone who thinks that data transfer is mainly one-way—from server to home—is ignoring the growing number of designers, writers, and engineers who work at home and transfer large CAD and desktop publishing files upstream.
- Interactive games—real interaction with other people in your town, but within the comfort and convenience of your own home—will bring new excitement to communities. This requires local bi-directional broadband connectivity.
- Super navigators that give you previews of programs on-request, not merely sorted listings, will make the new world of multi-channel digital video much more inviting. This requires interactivity.
- Last, because the excellence of the incumbent services makes it more of a long shot, is telephone service over cable. We will discuss how Personal Communications Services (PCS), a wireless telephony service, offers subscribers more than either wired or cellular telephony and how well the PCS application is matched to the capabilities of modern two-way cable plants.

In reality, however, the return path *resource* is still very much underutilized by the cable industry.

The Emergence of the Return Path

The design and operation of cable TV systems to date has been biased toward the forward path. This is because, until recently, the majority of the revenue opportunities for cable operators resulted from sending video signals downstream from the headend to the subscribers. The return path has played subsidiary roles: providing occasional pay-per-view requests from set-top converters, carrying video coverage of town council meetings from the town hall to satisfy a franchising requirement and, even more rarely, gathering information about the operational status of the plant equipment. As a result, cable systems have been designed to deliver one-way, downstream video programming, and the design of the return system has been something of an afterthought.

With flattening growth from traditional services and with competitive threats from telephone companies and from terrestrial and satellite microwave service providers, cable operators need to add new services to increase revenues. To provide most of these revenue-enhancing services, the one-way world of the cable TV operator needs to become a two-way world.

It is unfortunate, but understandable, that during the years of cable's rapid development, relatively little attention was given to the return system. Therefore, the foundation of knowledge and experience necessary for designing, building, and operating this part of the plant is only now beginning to become established and to be disseminated within the industry.

This book attempts to bring together the required information about the return system and about the applications that travel through a fully two-way network. We start by trying to dispel two myths that have become widely held: one that understates the problems of the return path, and a second that makes them appear too daunting. A reasonable balance is proposed between these two extremes, and—as a final motivation—the service revenue potential of the return path system is estimated.

Two Myths About Return Path Performance

Myth #1: "The return path is working today. It's a piece of cake."

Many cable operators who have an active return system are using it only for pay-per-view requests from set-top boxes. Since this application runs smoothly with relatively few difficulties in set-up, it would be easy to conclude that the addition of other interactive services would be a straightforward matter. Four factors make the favorable experience with two-way addressability somewhat misleading.

1. The upstream communication from set-top converters is by a "store and forward" protocol, which is less demanding than real-time communication.
2. The modulation scheme, frequency shift keying (FSK), is probably the most robust, but least bandwidth-efficient, of the digital methods.
3. The upstream data rate is very low.
4. Often there are no other services sharing the return band.

"Store and forward" means that a message to be transmitted upstream is stored in the set-top and is sent upstream by that unit only when the information is specifically requested by the headend equipment. Hence, the return system needs to accommodate only one set-top talking at a time. The transmit level of the individual set-top is established actively by headend equipment so that the signal will be strong enough to be received error-free. With no other applications sharing the return band, the converters can always be set to levels high enough to overcome any noise sources, since, if necessary, all of the RF power capacity of the return band can be allocated to this single transmission.

As a result, the pay-per-view system is able to operate in the presence of considerable noise. This allows the operator to combine the RF return signals from a large portion of the plant into one addressable system receiver, which is highly cost-effective. As we will see, when multiple services are operating simultaneously, with higher data rates and bandwidth efficiencies, the noise that would be present in all of these return connections is likely to cause detection difficulties. We will discuss in detail the two methods of attack on the noise problem: (a) reducing noise ingress into the system and (b) managing the return connections effectively.

Myth #2: "I've heard some real horror stories about ingress problems. The return path of cable TV systems will never work."

We will document a host of potential noise generators, we will show how those emissions can find their way into a cable system, and we will show how the very nature of the return system architecture causes the noise to accumulate. The fact that both the source and the entry point for much of this noise are inside the subscribers' homes—therefore largely outside the control of the cable operator—adds to the troublesomeness of the problem. Notwithstanding this, we will show how some of these sources can be controlled, how the entry of RF noise into the plant can be minimized (even though much of it may originate in homes), and how the optimal segmentation of the return system can be found to maximize cost-effectiveness. The key point is that **the return system *can* be made to work**, but it will require some new techniques and new technology, along with the organizational desire to make it work.

> *Once you get everybody to understand how they have to operate [the 5 to 40 MHz spectrum], it works extremely well. There's nothing you'll find that you can't take care of; sometimes it just takes a little longer than others. But it works. When customers see the result of that, they're awed by the speed and capability.*
>
> Tom Staniec, Vice President, Network Engineering for Time Warner's Road-Runner service.[3]

The Revenue Potential of the Return Path

So, if the return path is going to require effort and investment by the operator, how much revenue can it be expected to generate? We can estimate the income under different service scenarios. (We are looking only at the revenue line of the income statement, on the assumption that the operator will be able to determine the cost side.)[4] The revenue estimates are summarized in Table 1-1, and the method for estimating them is outlined in the discussion that follows.

Impulse pay-per-view

A return connection to the set-top converter permits subscribers to choose pay-per-view programs "on the spot," hence the name *impulse pay-per-view* (IPPV). A telephone return is possible, of course, but the cable return system is the logical path for these upstream request signals because there is no need for the installer to bring a telephone connection point close to the television set. The

3. *Multichannel News* (Chilton Company, New York), June 2, 1997, p30A.

4. A rule-of-thumb for the cost of RF and fiberoptic equipment for the return system is $7 per home passed (assuming 100 homes per mile of plant and 1000 homes passed per node). These items will be discussed in detail in the following chapters.

Table 1-1 Revenue projections for new services transported over the return path

	Monthly revenue per subscriber ($/mo)	Take-rate (%)	Annual revenue per 100,000 subs ($MM/yr)
Addressable set- top			
IPPV	15	60	10.8
Video navigator	1	75	0.9
Games			
Interactive	5	20	1.2
Cable modems			
Internet access	10	25	3.0
Telecommuting	60	10	7.2
Telephone			
PCS trunking	5	20	1.2
Cablephone	30	5	1.8

revenue estimate from the service is based on an average of three IPPV choices per month at a cost of $5 each.

Video navigators

As the number of channels increases, so does the need for some sort of two-way *interactive program guide* (IPG). This program proliferation is expected to explode as digital cable transmissions become more prevalent because digital compression will allow perhaps a dozen channels of entertainment per standard video channel bandwidth (6 MHz in the U.S.). IPGs make it easy for the viewer to find the best programs because they can do searches based on specific viewer preferences for categories, artists/teams, and times. As mentioned, the service can be enhanced significantly by the ability to access program previews, which can only be done interactively. The revenue from an IPG service is estimated at only $1 per month, but with a high subscriber take-rate.

Video games

Unlimited downloadable video game services, such as the Sega Channel, are already available for a monthly subscription fee on the order of $15. That service—which does not require an active return path—provides the latest games to be played in the home. As the software and hardware evolve, however, it is easy to envision fully interactive game services that would allow someone in one household to play against a friend or relative or even to have team events linking many homes. That would require low-latency return path signaling throughout the network. This would be well suited to servers located in the community's headend. The monthly revenue is estimated at $20 because—although this is a significant enhancement of the existing service—there is a limit on how much a household would be willing to spend for that service on a regular basis. At $20/month, however, as many as 20 percent of households could subscribe. While the return path could take credit only for the $5/month incremental revenue, interactivity would also attract new subscribers and retain existing ones.

Cable modems

It is estimated that more than 20 million U.S. homes access the Internet via personal computers, an increase of 12 percent in the first quarter of 1997 alone.[5] Because the signal is carried over twisted pair telephone lines, it is limited to data rates of tens of kilobits per second for standard telephony line cards. Coaxial cable access via cable modems raises that rate to megabits per second in each direction. The higher speed and the elimination of the need for extra telephone lines (or household disputes over use of a single phone line) should induce a majority of these households to switch to cable access services. Cable modems—unlike all telephony-based access systems—can be "on-line" for long periods of time without hogging system capacity. This is because the cable medium is a shared resource, like an office network, rather than a switched connection. This makes it practical, for instance, to have Internet servers in the home. In Table 1-1 we have accounted only for a $10/month *increment* in charges for the enhanced capabilities of a two-way cable modem, but we have assumed as many as 25 percent of the homes will subscribe.

In addition to the home PC user, an increasing number of professionals work out of their homes either as telecommuting employees or as independent

5. G.H. Arlen, *Information & Interactive Services Report* 18, no. 16 (April 25, 1997).

contractors. According to a study by FIND/SVP,[6] the US had 11.1 million telecommuting households in 1997, which represented a 15 percent annualized growth rate over the prior two years. Thirty-one percent of these households (3.4 million users) access the Internet regularly. Broadband access is expected to accelerate that usage rate, since it will enable videoconferencing and swift transfers of large-scale files for graphic artwork, computer-aided designs, and even video. Thus, it is not hard to foresee telecommuters connected to cable communications systems in 10 percent of all U.S. households. Telecommuting is likely to be more prevalent in certain regional markets than in others. The monthly rate for service should be comparable to business telephone rates.

Data service to commercial establishments is, of course, another revenue opportunity. This is especially true in the suburban United States, where businesses are often located along existing cable plant. We have *not* included an estimate of that revenue potential.

Telephone service

There are two ways for a cable TV operator to participate in the markets for telephone service: (a) as a direct competitor of other (wireline) local access phone companies or (b) as a supplier of signal distribution for another company that provides *personal communications services* (PCS) wireless telephony in the cable operator's service area.[7] (PCS trunking is explained in Chapter Eight.) Its attraction is that it allows the PCS operator to expand the area of coverage incrementally as subscriber traffic grows, while avoiding many of the difficulties in obtaining antenna sites. In the highly capital-intensive world of the PCS operator, the capability of cable to offer both *coverage* with *rapid deployment* is of particular importance.

Cable telephone revenues would be similar to residential phone rates, but the penetration rate is expected to be on the order of 5 percent. PCS penetration could be higher because it offers an attractive level of convenience. The revenue

6. *1997 U.S. Telecommuting Survey*, FIND/SVP, Inc., New York (July 1997).

7. The cable operator can be the PCS service provider, as well. In this discussion we are counting only the revenue from PCS transport and ignoring income from the PCS telephony service.

shown in Table 1-1 for PCS trunking includes only the transport revenue that would be paid to the cable operator by the PCS service provider.

Example

Using the information in Table 1-1, a cable operator with 100,000 subscribers who could provide addressable converters, interactive games, cable modems for home and telecommuting uses, and PCS trunking (but *not* cable telephone service) would receive an estimated $24 million in annual revenue for these additional services. That is the equivalent of a *system average* of over $20/month (for every one of the 100,000 households).

Summing-Up...

- The cable return system is a unique resource.
- It is a critical element for the future growth of the industry.
- It will not be easy to exploit, but it can be made to work.
- In many cases, the investment can be justified financially.

CHAPTER 2

Hybrid Fiber/Coax Cable TV Systems

This chapter will describe the basic elements of cable TV plant layout and equipment. The emphasis will be on hybrid fiber/coax cable TV networks, since that is the type of system over which all of the advanced services are expected to operate. Because all of the systems discussed in this book are two-way, we start by answering one obvious question.

How Do Both Signal Directions Flow Through a Single Cable?

First we need to describe how multi-channel signals are carried over cable systems. RF systems are *carrier* based, which means that a particular communication frequency is assigned for each communication channel and a pure tone at that frequency (the carrier) is caused to vary in some way by the information to be transmitted. For analog transmission of video on cable systems, it is the amplitude of the carrier that is varied, by the process called *amplitude modulation* (AM). This method has the obvious advantage of being compatible with the subscriber's TV receiver. All of the individual video channels are then combined in a *frequency division multiplex* (FDM), with the carriers separated usually 6 MHz apart (for NTSC systems). In the US, the forward channels are placed at frequencies above 52 MHz, while the frequencies between 5 and 40 MHz[1] are allocated to the return signals (Figure 2.1).

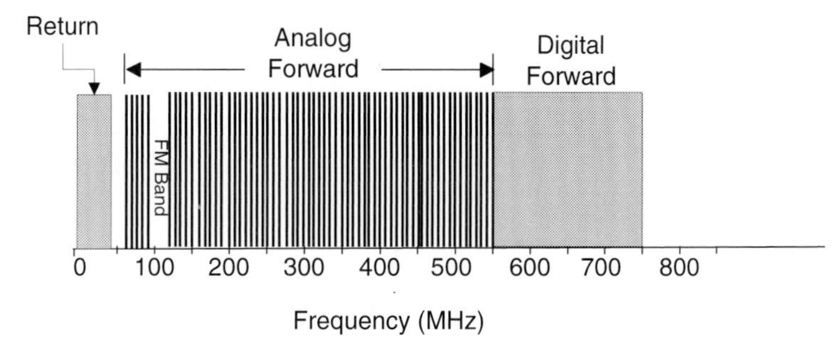

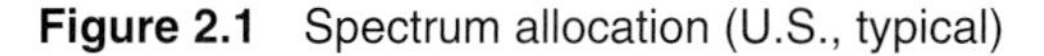

Figure 2.1 Spectrum allocation (U.S., typical)

At various points in the cable system, amplifier units are inserted to restore the signal strength of both the forward and return signals. Since amplifier circuits are inherently unidirectional, the amplifier unit must first separate the signals flowing in the two directions. This division is done as shown in Figure 2.2 by circuits known as *diplexers*,[2] which have the frequency response shown in Figure 2.3. After the diplexing, each signal is amplified and again connected to the coax cable (through an identical diplexer).

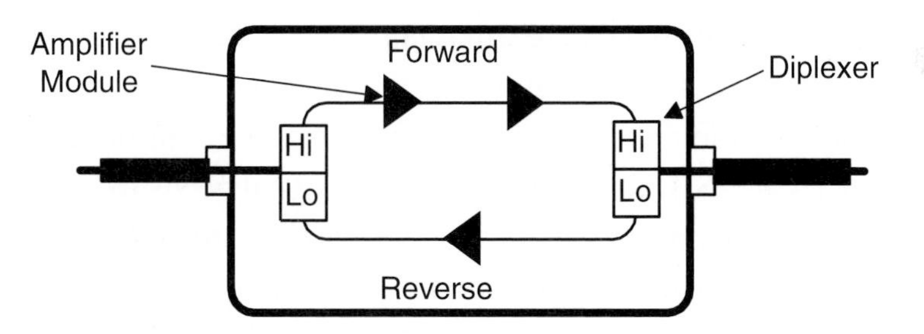

Figure 2.2 Amplifier block diagram

1. The 40/52 split [the notation indicates high-end of return/low-end of forward] is used throughout North, Central, and South America and in many systems in the Far East (China, Korea, Philippines, Thailand, and Singapore), with some systems having the return band end at 42 MHz and the forward band start at 54 MHz. Prior to 1994 the North American return band ended at 30 MHz. Other common frequency splits are Australia 65/85; Japan and New Zealand 55/70; India, Malta, and Eastern Europe 30/48; Western Europe, Ireland, and the United Kingdom 65/85. Some systems do not use the band from 5 to 10 MHz due to concerns about noise ingress and signal rolloff.

2. Communication engineers outside the cable TV industry often use the term *duplexers* for circuits that perform this function.

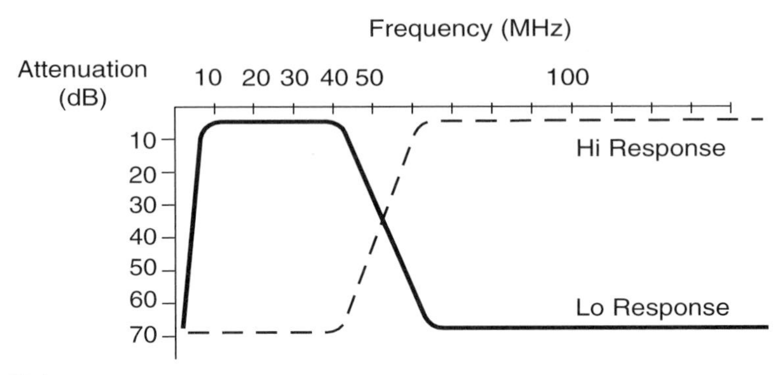

Figure 2.3 Diplexer response

Elements of Cable TV Systems

Forward path

Figure 2.4 shows a simple HFC system in block diagram form so that we can define some common terms. The *headend* contains the equipment that receives the television signals from satellites, terrestrial microwave, cable, and over-the-air sources and arranges them for transmission out onto the cable plant. The satellite microwave signals are often *multiplexes* consisting of several programs in digital form that were combined for *uplink* and *downlink* using one satellite transponder. If the received programming is to be sent out over the cable plant in analog (AM-VSB) form, the microwave signal needs to be demodulated down to digital baseband, decrypted, demultiplexed into individual programs, and converted to baseband TV audio and video signals. These functions are done in an *integrated receiver/decoder* (IRD). The single-program baseband output from the IRD is then scrambled for conditional access in a *modulating video processor* (MVP), which puts out a composite video/audio signal at a standard *intermediate frequency* (IF).[3] The scrambled IF signal is then put onto its assigned *radio frequency* (RF) transmission channel in an *upconverter.*

A digital satellite multiplex that is intended to be transmitted digitally (as a multiplex) over the cable system is received by an *integrated receiver/*

3. Usually with picture carrier at 45.75 MHz in NTSC countries, 38.0 MHz in China, and 38.9 MHz elsewhere. To facilitate the next frequency conversion, the audio carrier is *below* the picture carrier.

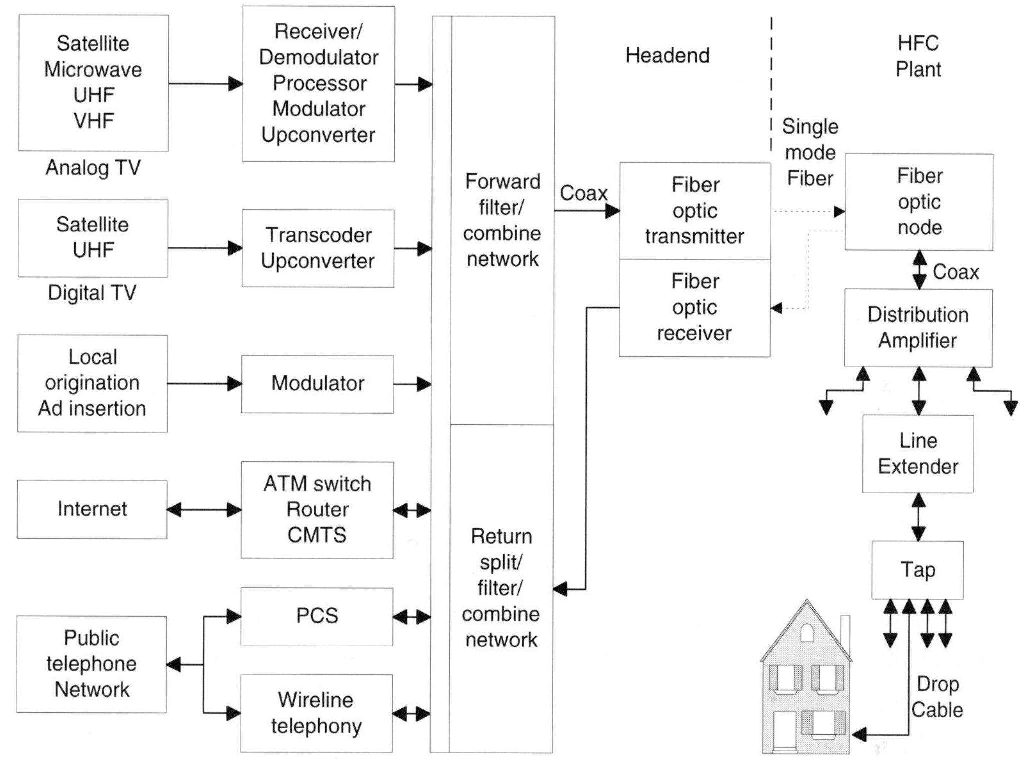

Figure 2.4 HFC terminology

transcoder (IRT), which demodulates it from L-band, decrypts it, and then QAM *modulates* it (as will be discussed in the next chapter) onto an IF carrier so that the digital signal will be compatible with set-top converter systems. An upconverter is used to convert the signal from IF to the assigned RF.

Over-the-air signals are sent either to *processors*, which downconvert the incoming carrier and upconvert to cable RF, or to separate *demodulators* and *modulators*.

Outputs from the upconverters and modulators are combined to form the full complement of signals to be sent to a given neighborhood. This set is amplified and applied to a *fiberoptic transmitter* that usually incorporates a semiconductor laser specifically designed for analog transmission. The electronic RF signal modulates the optical output power of the laser, thus converting the AM electronic signal into an AM optical one and putting it onto one of the optical fibers contained in a fiberoptic cable.[4]

Optical transmission of cable TV signals can be done essentially only with *singlemode optical fiber*, which has an attenuation of approximately 0.4 dB/km at 1310 nm optical wavelength and 0.25 dB/km at 1550 nm.[5] The attenuation of optical fiber is constant over reasonable temperature ranges and is independent of RF frequency. Typical fiber cables contain 12 to 144 fibers.

At the other end of the fiber is a *fiberoptic node* station (Figure 2.5), usually mounted outdoors, either on the cable strand or in a pedestal cabinet on the ground. The node receives the optical signal, converts it to electronic form, amplifies it, and begins the distribution over the coaxial cable network. In general, there are three or four coax outputs from the node. Large diameter (0.625" to 0.875") coax cable is used in this portion of the network, with a solid aluminum outer conductor and a solid copper or copper-plated steel center conductor. Its attenuation is on the order of 0.015 dB/ft at 750 MHz and 0.003 dB/ft at 40 MHz (approximate square root of frequency dependence). This attenuation is temperature-dependent, as well, increasing by approximately 0.11%/°F (0.2%/°C), due to the elongation of the cable.

In many designs, the cable span from the node to the next amplifier—a *distribution amplifier*—does not feed subscriber homes directly (this is referred to as *express* cable). The distribution amplifier unit amplifies the RF signal and puts it out on two to four output cables. These cables serve two purposes: they contain *taps* to supply signals to the homes passed, and they carry the signal outward to other distribution amplifiers or to single-output *line extender* amplifiers. Typical HFC systems will have RF cascades of 4 to 6 distribution and line extender amplifiers after the node, several of which will have automatic gain controls (AGC) to compensate for the change in cable attenuation and frequency response due to temperature.

4. Optical splitters can be used at the headend to send the same composite optical signal to several locations. This reduces the initial investment in optical transmitters, but at the expense of narrowcasting capability.

5. See Appendix D for a discussion of decibels (dB).

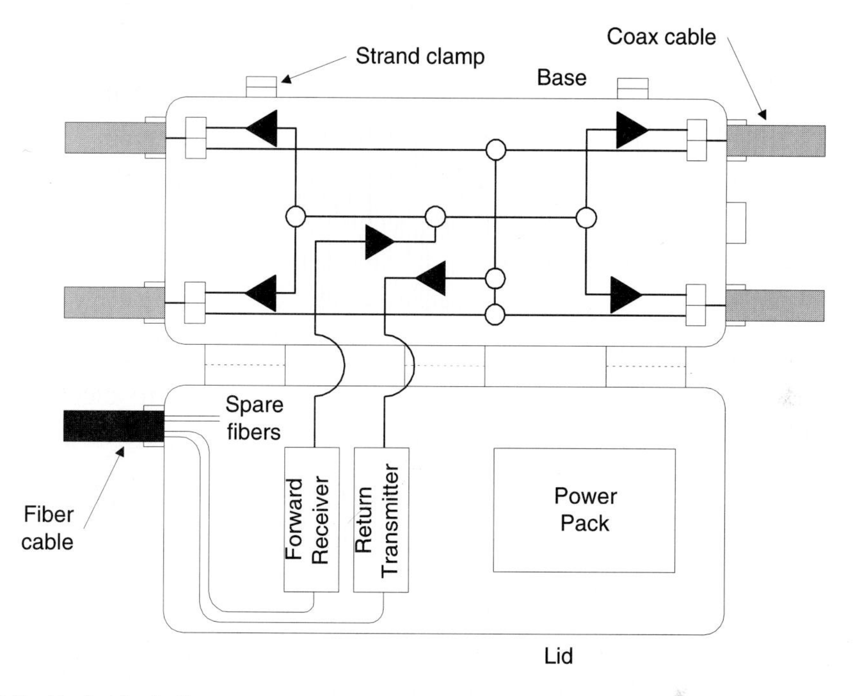

Figure 2.5 Node block diagram

As we will discuss in Chapter Eleven and in Appendix B, the cable plant is designed so that the forward output of each amplifier in the cascade is the same (*"unity gain"* design). Because the cable has higher attenuation at high frequencies, the amplifier output is *tilted* upward by 8 to 12 dB. This has the side benefit of reducing the intermodulation distortions between the various channel signals by reducing the total power through the active components of the amplifier. As forward digital services are added to the analog video programmming, they are usually inserted at frequencies above the analog ones. Since the carrier-to-noise requirements for the digital transmissions are not as demanding as for the analog, the power of the digital signal is, in general, reduced by a fixed amount (6 to 10 dB), which reduces the amount of additional power loading on the amplifiers (Figure 2.6).

The taps mount directly in-line with the cable and consist of a housing that contains a *directional coupler* component and *power splitters* (Figure 2.7). The directional coupler diverts a specified amount of the input signal, and the split-

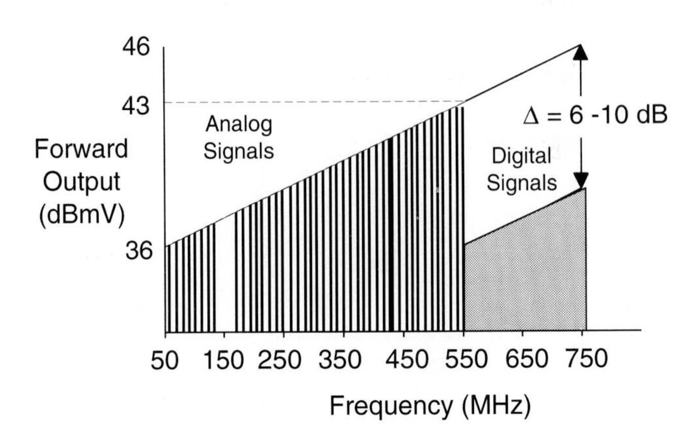

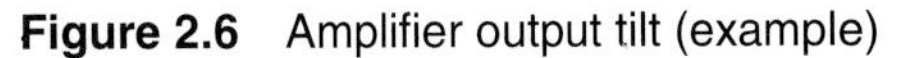
Figure 2.6 Amplifier output tilt (example)

ters divide that signal between the number of tap ports in the unit (2, 4, or 8, usually). Taps are denoted by a *tap value*, the ratio (in dB) between the signal at the tap and the input signal. Common values are 4 to 29 dB, in 3 dB steps. Tap loss is nominally not frequency- or temperature-dependent. The through signal on the distribution cable is attenuated by the loss due to the power tap-off plus 1–2dB of intrinsic losses.

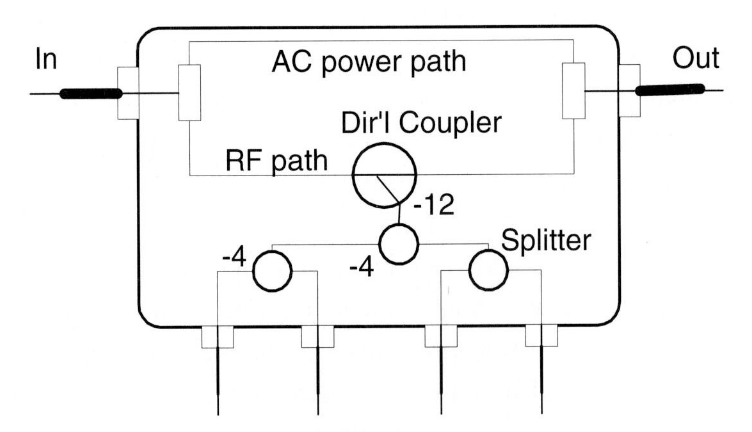

Figure 2.7 Block diagram of four-way tap (20 dB tap value)

The subscriber's home is served by approximately 75–150ft of *drop cable* of smaller diameter coax (usually RG-11 in new installations)[6] that carries the signals from the tap to the side of the house. RG-11 has an attenuation of .038 dB/ft at 750 MHz, and 0.008 dB/ft at 40 MHz. The drop connects to the *ground block* on the side of the house, which provides grounding of the cable shield (usually required by local building codes) and provides a 75-ohm penetration through the outside wall. Typically in the U.S., the cable operator also provides in-home cabling, at least to the first TV set.

One of the critical plant design choices for the operator is the number of homes passed per node. As will be discussed in Chapter Eight, this choice affects the number of services that can be offered and the number of subscribers that can be provided with each service. Systems are being designed today with 350 to 2000 homes passed per node. In general, designs calling for 2000-home nodes include provisions for a straightforward subdivision into four 500-home segments in the future, as markets develop.

Return path

The identical coaxial cable path is used for return as for forward. A return signal from the home goes up the drop, through the tap, inserts into the distribution cable via the tap's directional coupler, and passes through amplifiers. When the return signal gets to the node station (Figure 2.5), it is diplexed, amplified, and sent to a return laser so that the signal can be transmitted up to the headend over optical fiber. (In nearly all cases this is not the same fiber as the one used for the forward signal.) At the headend, the optical signal is photodetected and converted back to electrical. It is then split to feed the receivers for the various return service applications.

Powering

Alternating current (AC) is distributed on the coax cable network to power the nodes and amplifiers. The power is drawn locally from the electric utility, hence it is always at the same frequency as the utility (60 Hz, except in parts of Europe, Australia, and New Zealand where 50 Hz is standard). The utility power is regulated by *ferro-resonant transformers*, which are very rugged and stable,

6. The RG-6, RG-11, and RG-59 nomenclature, which comes from military procurement, is slowly being replaced by the terms 6-series, 11-series, and 59-series.

but which put out a square waveform. As attention to network availability performance has increased, it has become more and more common for the network power supplies to be backed up by batteries that are always on trickle charge.

In each node and amplifier, the AC power is separated from the RF signals by a low-pass filter and sent to an on-board power pack that converts it to +24 volts DC, the industry-standard supply voltage for hybrid amplifier components. The AC is then re-inserted into the output ports, if required to power other amplifiers downstream.

As we will see in Chapter Eight, many telephony applications require power to be supplied to the home, as well, which adds considerable complexity to the powering design and to the network components, such as taps. One aspect of this is that many operators are moving to 90 VAC from the previous standard operating voltage of 60 VAC so that telephone set powering can be accomplished at more practical currents and at lower transmission losses.

Hybrid Fiber/Coax Networks

The preceding discussion assumed the use of fiberoptic links in both the forward and return systems. In systems being implemented today, this is almost universally the case, but it represents a near-revolutionary advance in cable TV networks that became the standard for the industry only a short time ago. This distribution method is based on the use of semiconductor lasers and singlemode optical fibers to carry cable TV signals in their native, analog form, which was first demonstrated commercially only in 1990.[7] It took less than three years from that first demonstration for the cable TV industry throughout the world to appreciate the benefits of fiberoptics and to incorporate "AM fiber" technology into a totally new system architecture. That architecture—now called *hybrid fiber/coax*

7. As we understand it, the first prototype field-deployable laser transmitter for AM cable TV was given to Time Warner (American Television Communications) by personnel from Lucent (AT&T, at the time) for evaluation in their laboratory. Within days it was put into a service trial in Florida to back-up an *AM microwave link* (AML) that was beset by weather-related fades. After enduring two weeks more of rain fades, Time Warner made the laser link the primary path, with the AML as back-up. The system remained that way (with the prototype unit in service) for about 18 months before being replaced with fully commercial fiberoptic equipment. Lucent presented the laser to Jim Chiddix of Time Warner for his role in spurring the technology.

or HFC—brought many efficiencies to the cable TV plant. Most fundamentally, however, the HFC architecture made it possible to think of delivering *narrowcast* services, targeted to specific neighborhoods, and *interactive* services, targeted to individual homes. This is because HFC replaced the traditional *tree-and-branch* architecture[8] (Figure 2.8)—which in essence connected all subscribers in one long chain—with a system of smaller clusters (Figure 2.9), each tied directly back to the headend.

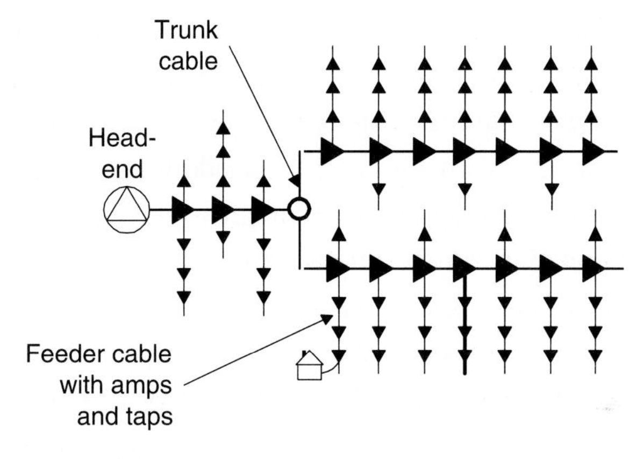

Figure 2.8 Tree-and-branch system

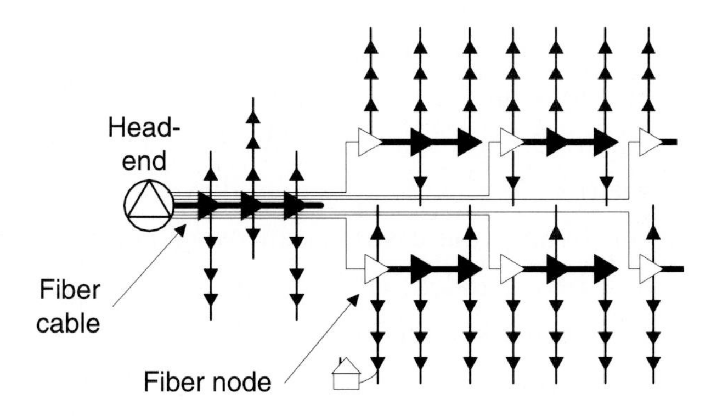

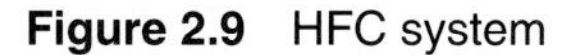
Figure 2.9 HFC system

8. Sometimes called trunk-and-feeder, since the central amplifier cascade is called a *trunk* and the branches *feed* subscribers' homes.

In Figure 2.8, one can see that all of the subscribers receive the same programming (except that certain premium programs can be blocked by in-line trapping devices in the subscriber's drop). The clear strong point of that network is that it delivers large numbers of video channels[9] by using high bandwidth coaxial cable all the way to the subscriber's home. Even the best coaxial trunk cable attenuates a 550 MHz (77-channel) signal at 0.01 dB/ft, however, which means that RF amplifiers with 27 dB gain are needed every half-mile (0.8 km). Thus, the trunk needs a cascade of many of these amplifiers (a single 15-mile trunk requires 30 of these amplifiers in series). Hybrid fiber/coax builds upon the same broadband coax delivery system to the home, but uses fiberoptic technology to eliminate the trunk cascade and to segment the system, as can be seen by comparing Figure 2.9 with Figure 2.8. In addition to enabling targeted services, HFC shortens the average number of active devices between headend and subscriber. The inherent service reliability of the network increases both because the number of actives in series is reduced and because the failure of any one component affects fewer subscribers. The fiberoptic components themselves and the fiber cable have shown very high reliability.[10]

One additional benefit of HFC is that the system is readily extensible, meaning that the network can be further subdivided to increase bandwidth per subscriber without major re-investment. Note that the bandwidth allocated to upstream communications on the return path is much less than that devoted to downstream (see Figure 2.1, for example). While many of the planned digital services are asymmetrical (e.g., video program downstream, with only requests for programming upstream), there still is an expectation that return network capacity will need to be expanded. HFC makes this relatively straightforward for the operator. As the market for more highly targeted services grows, the cable operator can provide the necessary bandwidth by making only modest incremental investments in plant, as will be described in Chapter Eight. This is a major distinction between HFC delivery of broadband services and switched

9. Tree-and-branch systems with up to 77 channels have been built in the U.S.

10. See Chapter Seven.

(telephone-like) digital delivery, where essentially all the capacity for hypothetical future services must be built in at the beginning.

Summing-Up...

- The elements of a cable system headend and distribution plant were described.
- The return path shares the distribution elements with the forward.
- AC power is also generally distributed on the same coaxial cable.
- HFC network architectures permit narrowcast programming by a segmentation of the subscriber base that can be further subdivided as the market demand grows.

CHAPTER 3

Digital Applications on Cable Systems

All of the signals traveling over the return path will be digital. Therefore, before we start our detailed discussion of return path design, operation, and performance, we need to establish a foundation in digital technology. This chapter begins with a general explanation of how digital signals are generated from analog information, and explains how they are carried on cable systems. We will compare the performance of a variety of transmission schemes. Finally, we will discuss several techniques that are commonly used for communicating over noisy channels.

The purpose of this chapter is to make the remainder of the chapters understandable, but in the interest of brevity, much of the material will be treated somewhat superficially. It is fortunate that many excellent references that present the details on this subject are available to the reader whose curiosity has been raised.[1]

Again, we start with a question.

1. Bernard Sklar, *Digital Communications: Fundamentals and Applications* (Englewood Cliffs, NJ: Prentice Hall, 1988).
George R. Cooper and Clare D. McGillem, *Modern Communications and Spread Spectrum* (New York: McGraw-Hill, 1986).
Edward A. Lee and David G. Messerschmitt, *Digital Communication* (Boston: Kluwer Academic Publishers, 1988).

Why Digital?

Digital technology has become so much a part of our everyday life that it almost seems silly to ask this question. On the other hand, the natural world is an analog place: time flies—it doesn't hop; people talk endlessly (some people, at least); and the passing scene flows by. So why are we wearing digital watches, why are we talking through digital telephone systems, and why are we entering an era of digital television? The answer (as everyone knows) relates to the achievement of incredibly high circuit densities in digital integrated circuit technology. This has allowed us to employ digital techniques to deal with formerly difficult analog phenomena using essentially brute force computation. An example is the digital spectrum analyzer that is used throughout our industry, which uses fast Fourier transforms rather than finely tuned analog circuits, to determine the spectral composition of an RF waveform. A second important aspect is that by reducing a signal transmission to merely ones and zeros there is the prospect for error-free communications. This is because digital communication requires only that a *decision circuit* determine whether the incoming bit is a one or a zero. Once the decision is made, the bit can be *regenerated* (reshaped, retimed, and amplified) for re-transmission with no loss. In the analog world, on the other hand, amplification is the only tool available, unavoidably causing some degree of signal degradation at each stage.

Generating Digital Representations of Analog Signals

Two processes are employed to convert an analog signal into digital form: sampling and quantization. Two additional processes, coding and modulation, are critical to transmitting that digital information.

Sampling

An analog signal, such as your voice going into a telephone, is a continuous entity. The digital representation of that analog waveform is a series of discrete numbers corresponding to the *values* of that signal measured at a sequence of uniform time intervals. This is called *sampling* the signal.

There is a well-defined way to determine how many samples are required to fully characterize the signal-at-hand. It is given by *Nyquist's theorem*, which states that if a signal has frequency components up to a maximum frequency f_{max}, then the sampling rate f_s must be at least twice f_{max} (that is, $f_s \geq 2f_{max}$). This can be understood for the simple case of a pure sinewave (Figure 3.1). We can see that two measurements per cycle of the wave are sufficient to distinguish that signal from any other signals of *lower* frequency.

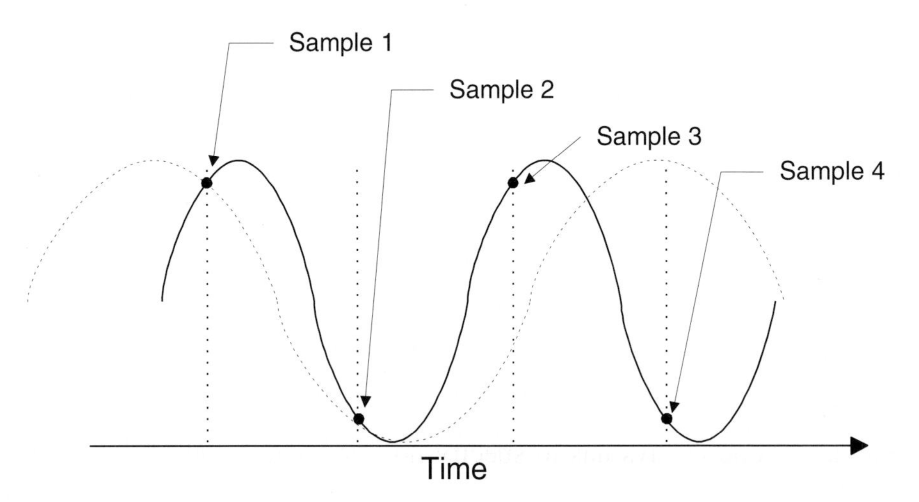

Figure 3.1 Nyquist sampling

Referring to Figure 3.2, we can see that measurements at this sampling rate will not distinguish the sinewave from another at *higher* frequency. This potential confusion is called *aliasing*.[2] It is aliasing that causes motion artifacts in movies, such as wheels appearing to rotate backward, because the sampling rate (pictures per second) is too low to faithfully reproduce high-speed events. The implication for digital transmission is that if a signal is going to be sampled at the rate f_s, then it must first be low-pass filtered at $f_s/2$.

Quantization

By sampling, we have turned the continuous analog signal into a series of numbers corresponding to measurements of that signal at intervals of time. It is necessary, however, to specify the precision of that measurement (or, more prop-

2. The term refers to the fact that one digital object has two analog "names."

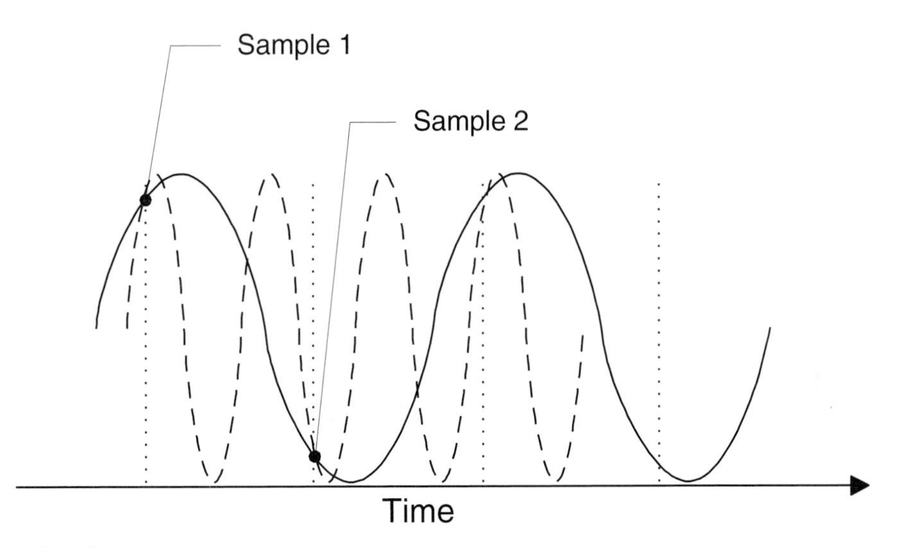

Figure 3.2 Aliasing

erly, the precision of the number we record for that measurement). In analog measurements, one always has to specify how many decimal places are being used; in digital parlance, this equates to saying how many bits are contained in the binary version of that number.

Accordingly, we allocate a certain number of bits (called a *word*) to each number in the series representation of our waveform. If, for example, we are using 8-bit representations, then the signal amplitude measurement is accurate to one part in $2^8 = 256$. This is equivalent to better than 1 percent accuracy on the average. This process of measuring in 1-bit steps is called *quantization* and the measurement increments are referred to as *quantization steps*.

Error generated by the digitization process

In essence we have put an amplitude-time mesh on the waveform with the amplitude on a $\log_2$ scale (Figure 3.3), and we have converted the signal into a *time series* or list of the amplitude cells through which it passes.

It is clear that when we sample and quantize, we are consciously "losing" some of the original analog information. The time series of numbers generated in the analog-to-digital process is not, strictly speaking, *identical* to the analog waveform. This means that the design engineer needs to find the right balance

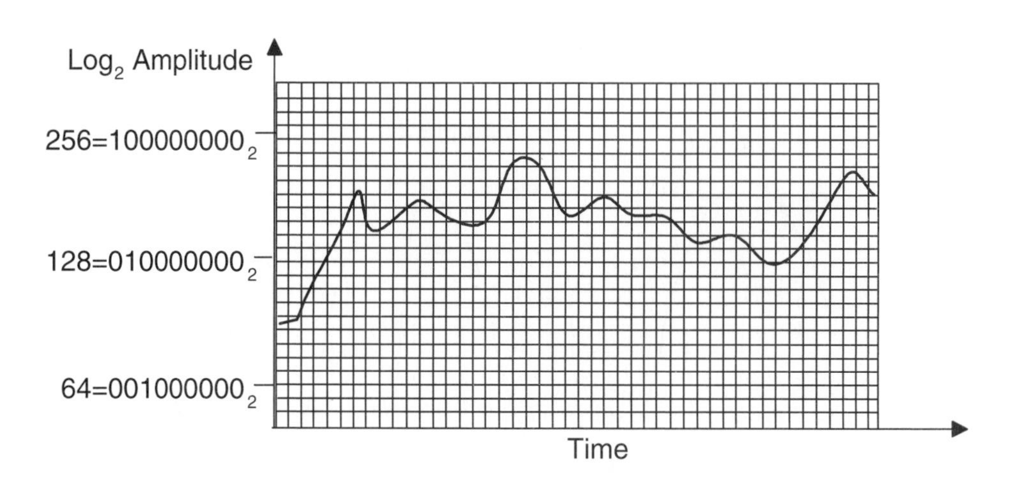

Figure 3.3 Quantization

between fidelity (which entails higher numbers of samples and more bits per sample) and economy (which calls for the minimum number of bits per second) when choosing the sampling rate and the word size. This is done by determining the *signal to noise ratio* (SNR)[3] of the analog-to-digital conversion process. This is given by the formula

$$SNR = 6.02N + 10*\log(f_S/2B) + 1.76,$$

where N is the word size (the number of bits in a sample), f_S is the sampling rate, and B is the low-pass filtered bandwidth of the analog signal.

The first term comes from the quantization accuracy of 2^N:1, since $10*\log[(2^N/1)^2] = 2 \times N \times 10*\log(2) = 6.02N$. The second term relates to the sampling process. It is zero when sampling is at exactly the Nyquist rate, but shows the benefit that is gained by *oversampling*. The final term accounts for the fact that the amount of error is distributed uniformly over the range of ±1/2 of the least significant quantization bit.[4]

3. Since most of the discussion in this book relates to carrier-based transmissions, we generally characterize the channel by a carrier to noise ratio (C/N). On the other hand, the term SNR is used when we are discussing the baseband signals (essentially the ones and zeros prior to modulation or after detection).

As a result, *only by oversampling* can an 8-bit digital transmission have an SNR better than 50 dB, or a 10-bit system better than 62 dB. This becomes important when video signals are transported digitally over a portion of a distribution network (Chapter Twelve).

Source coding

We now have a binary data stream that is a faithful representation of the analog information we want to transmit. It is often more efficient to transmit this information as *groups* of bits, referred to as *symbols*, rather than as individual bits. (We will explain how this is accomplished in the next section.) In a 4-bit-per-symbol system, there would be 16 different symbols, for instance *A* through *P* with *A* corresponding to 0001 and *P* to 1111. The advantage of such a scheme is that only *one* symbol needs to be transmitted and detected, rather than *four* bits. *Coding*, as this process of bit grouping is called, can be likened to verbal communication, where we speak in whole words rather than spelling out the letters. Coding of the data into symbols is referred to as *source coding*, to distinguish it from *channel coding*, which is the addition of bits for error correction, which we will discuss at the end of this chapter.

Transmitting Digital Signals on Cable Systems

We think of digital signals as a series of 0s and 1s that correspond to "off" and "on" states of transistors. In fiberoptic systems used for telephone trunking, this is essentially how they are transmitted. A laser is caused to turn on for a 1 and off for a 0. Thus the transmission consists of a series of flashes. The term for this type of laser modulation is *on-off keying* (OOK).[5] This *baseband digital* signal is often a composite of many lower rate data streams that have been time-interleaved in what is called a *time division multiplex* (TDM).

4. This term is 10*log[(1/2)/(1/3)] = 10*log[3/2] = 1.76. This is the log ratio of the variance (in power) of a sine wave (1/2) to that of noise with a uniform power density (which is the usual assumption for the noise characteristic for an analog-to-digital conversion).

5. The use of the word "keying" for modulation is, of course, a throw-back to the original means of electronic communication, Morse code.

Within cable systems, however, the digital signals are transmitted on modulated carriers, in a *frequency* division multiplex, just as the forward analog video ones are.[6] This means that the digital information must be transformed from its baseband state by modulation. The process is similar to the modulation of baseband video, but the modulation techniques are quite different. In fact, the first of these differences is that there are a *variety* of digital modulation schemes used, unlike the single AM technique employed for analog video. This is because there is no one "best" choice for digital modulation—unlike analog video transmission, where AM is the obvious choice because it keeps the signal compatible with the hundreds of millions of TV receivers fed by the cable system. The reason for this variety can be understood by examining some hypothetical system design questions.

- How data intensive is the application? If a high data rate is to be transmitted in a limited RF band, then high *spectrum or bandwidth efficiency* is required.
- How much noise, interference, or reflection must be overcome in the communication channel? If the channel is expected to be very noisy, then high *robustness* is required. To further complicate the subject, one modulation type may be more robust than another against a particular kind of channel impairment, such as reflection, but less resistant to another, such as group delay.
- How much cost can be allocated to transmitters and receivers? The equipment for the more spectrum-efficient schemes will be more complex, hence more expensive. Note that for the return path, the cost of the transmitter is critical since there is one per subscriber. Receivers, on the other hand, can be shared among groups of subscribers. Also as we will see, higher spectrum efficiency puts additional requirements on the performance of the return plant, which will impact system equipment and maintenance costs, as well.

6. This was discussed in Chapter Two.

The answers to these and other questions force trade-offs to be made between efficiency, robustness, and cost. Thus, the choice of modulation scheme depends on the relative weights specifically assigned to the various performance requirements. Hence, it is not surprising that different products for different applications will call for different modulation techniques.

Modulation Techniques

Three parameters define a carrier—its phase, frequency, and amplitude—and modulation methods based on varying each one have been developed and implemented. We will portray each of them with a *state diagram*, which shows the positions of the bits or symbols in the phase plane, the frequency spectrum or the amplitude axes, as appropriate. Generalized transmission of binary (uncoded) information is shown schematically in Figure 3.4.

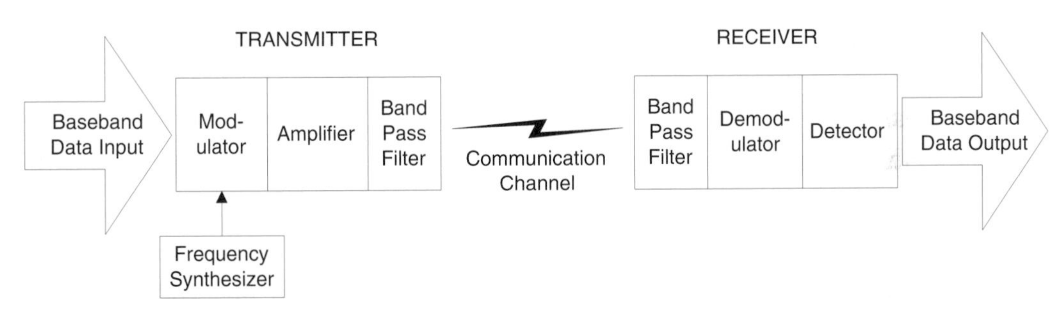

Figure 3.4 Binary communication block diagram

Binary frequency modulation

In *frequency shift keying* (FSK), the carrier frequency is shifted by each data bit. For example, if f_c is the carrier frequency, a 1 would raise the frequency to $f_c + \Delta f/2$ for one bit period, and a 0 would lower it to $f_c - \Delta f/2$. Figure 3.5 shows a particular type of FSK, in which the frequency shift Δf is chosen so that the phases match at the bit-period boundaries. This type of FSK has the advantage of containing more of the transmitter power within the transmission band since discontinuities are minimized. This means that the transmission is more efficient and the filtering requirement is less stringent.

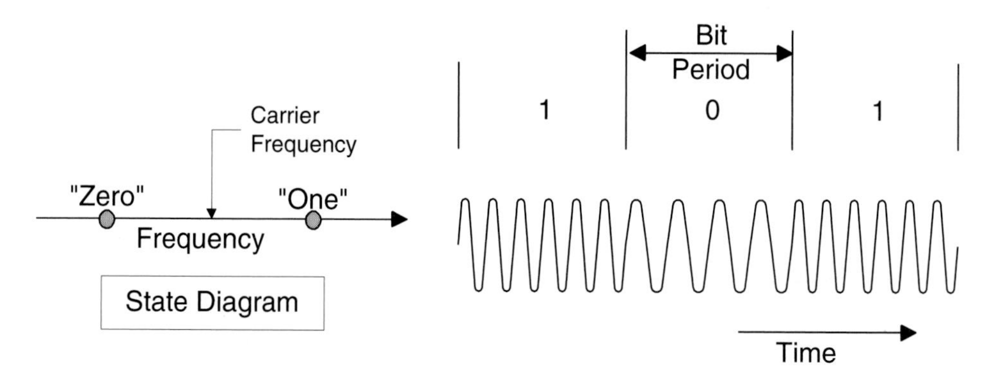

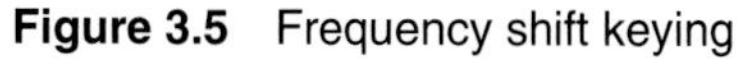
Figure 3.5 Frequency shift keying

Binary phase modulation

In *bi-phase shift keying* (BPSK), a 1 is represented by advancing the phase of the carrier by 90° past some reference for one bit period, and a 0 by retarding by 90° for one period (Figure 3.6).

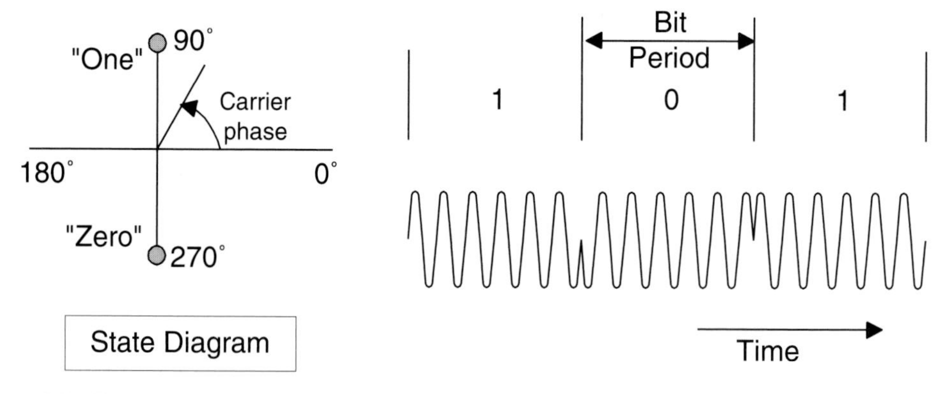

Figure 3.6 Bi-phase shift keying

Binary amplitude modulation

As expected, binary *amplitude shift keying* (ASK) modulates the carrier amplitude "positively" for a 1 and "negatively" for a 0. But what is meant by a negative carrier amplitude? The answer is: one with a reversed phase. Thus, this form of binary ASK is conceptually the same as PSK.

M-ary modulation

It is possible to establish more than one pair of states per axis, such as by using a phase plane divided eight ways,[7] as in Figure 3.7a, or by dividing an amplitude four ways, as in Figure 3.7b. Source coding is being utilized here, since each state now corresponds to a *symbol* of more than one bit. In particular, Figure 3.7 shows the use of so-called *Gray codes*. This class of codes has the property that adjacent symbols differ from each other in only one bit-position. Since errors made by the decision circuit in any detector are likely to be confusions between adjacent symbols, this reduces the resultant number of errored bits to one.

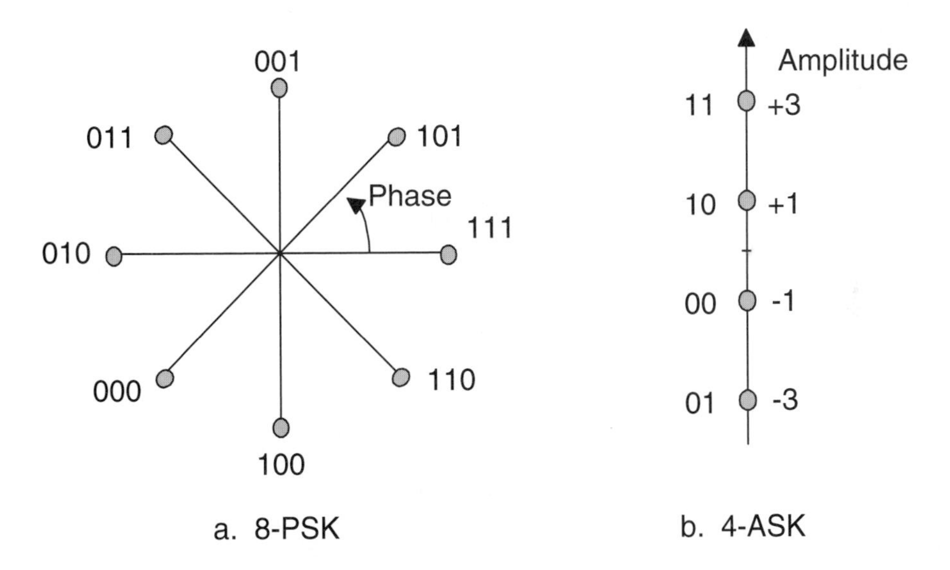

Figure 3.7 M-ary modulation examples

Quadrature modulation

So far the state diagrams we have used could have been single axes, rather than planes. It is possible to double the information-carrying capacity of a communication channel, with relatively little hardware cost, by using both axes in the plane through *quadrature modulation* techniques. Two carriers differing in

7. This modulation scheme (8-PSK) was used for 4800 bit/sec telephony modems.

phase by 90° are generated from one signal source, and each is modulated independently, as shown schematically in Figure 3.8. The two carriers are generally denoted I and Q, for in-phase and quadrature. Because the two carriers are mathematically orthogonal to each other, they do not interact. Hence, they can be combined for transport. At the detector, the two information streams can be demodulated independently. After demodulation, the I and Q bit streams are combined to recreate the original symbols.

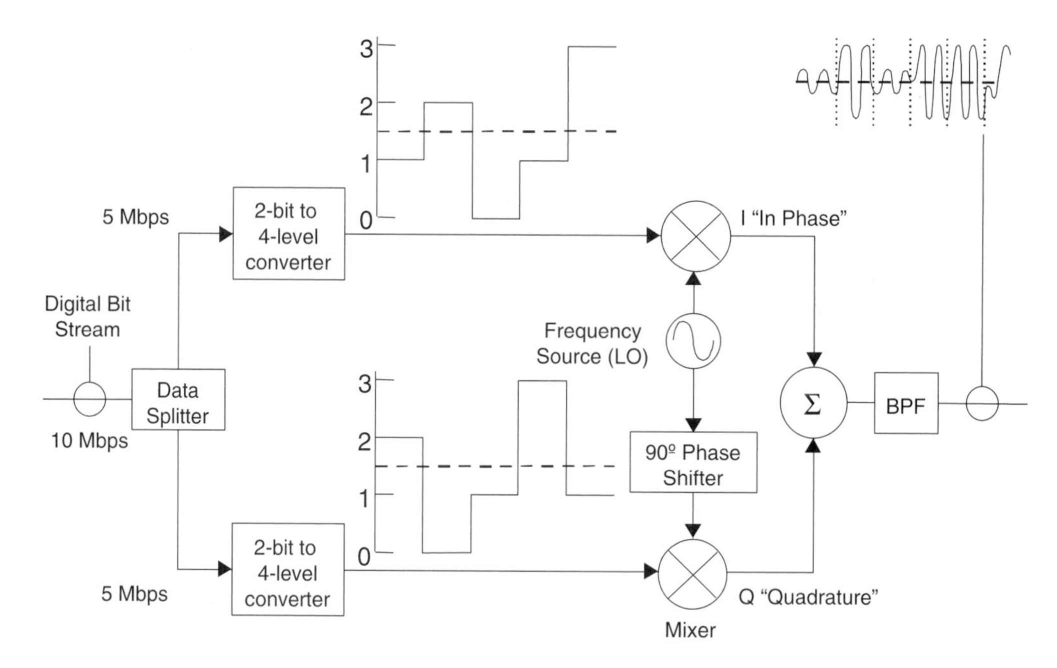

Figure 3.8 Quadrature modulation block diagram (16-QAM) BPF = bandpass filter

The simplest and most common quadrature modulation scheme—and one that is likely to be used often for return path applications—is *quadrature phase shift keying* (QPSK), which establishes four states by performing independent BPSK modulations on each of the I- and Q-axes (Figure 3.9).

A second quadrature modulation that is being used in cable systems is *quadrature amplitude modulation* (M-QAM). In M-QAM there are $\sqrt{M}$ amplitude states on each of the I- and Q-axes, which results in a *constellation* of M states representing symbols consisting of $\log_2 M$ bits. 16-QAM (Figure 3.10) is being

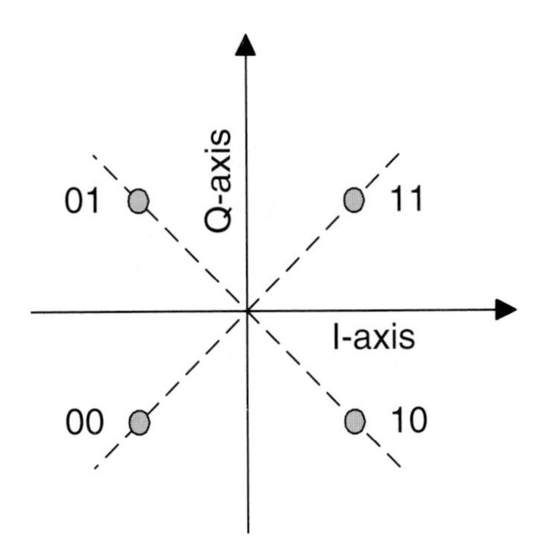

Figure 3.9 Quadrature phase shift keying state diagram

used in the return path, while 64-QAM and 256-QAM are being deployed in the forward path. Twisted-pair telephone modems have used 16-QAM at 9600 bit/sec speeds and 64- or 128-QAM at 14.4 kb/sec. Note that QPSK is functionally equivalent to 4-QAM.

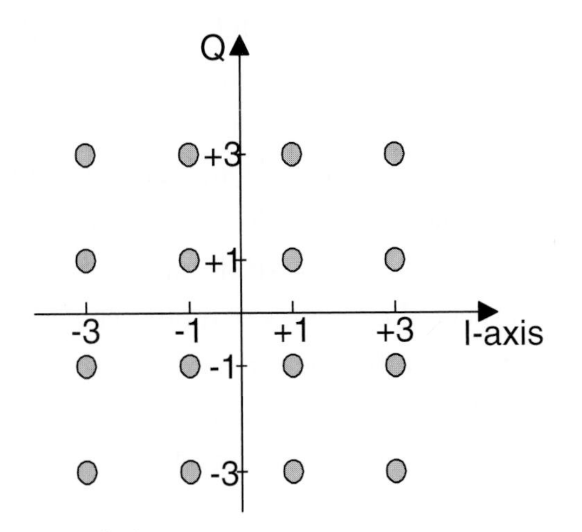

Figure 3.10 16-QAM constellation

Performance Comparisons

Up to this point we have dealt with the "what" and the "how" of digital transmission. The remainder of this chapter deals with the *quality* of that transmission. This will set the stage for much of the material in the following chapters, which deals with ways to ensure that the transmission quality is satisfactory. The two principal engineering yardsticks used to compare the various modulation schemes are the *spectrum efficiency*, which indicates the number of bits transmitted per unit of modulation bandwidth,[8] and the *energy per bit* (at a stated transmission accuracy). From the viewpoint of a cable system operator, the first quantity measures how many signals can be squeezed into a limited return bandwidth, and the second indicates how high the system carrier-to-noise needs to be. As mentioned previously, there are other important characteristics such as cost and complexity.

Bits per second per hertz

Spectral efficiency is determined by dividing the bit rate (the product of bits/symbol times symbols/sec)[9] by the modulation bandwidth. The modulation bandwidth is primarily a function of the filter design in the modulator and detector. A starting assumption for the *binary* modulation schemes is that the modulation bandwidth equals one bit/sec per Hz.[10] Quadrature techniques do not add to the bandwidth requirement, therefore QPSK has twice the spectral efficiency of BPSK. Based on these approximations, M-PSK and M-ASK, with a symbol size of $q = \log_2 M$ bits, have a spectral efficiency of q bits/sec per Hz. M-QAM has an efficiency of q bits/sec/Hz, as well, since it transmits $q = \log_2 M$ bits on every cycle of the carrier. M-FSK, on the other hand has an efficiency of 1/q, since each new state requires additional bandwidth. As examples, QPSK, which is

8. In units of bits/sec per hertz, but often shortened to "bits per hertz."

9. The term "baud rate" is sometimes used to denote symbols/second.

10. This is true for ideally shaped bit pulses. If, for instance, the pulse shape is rectangular, the spectral efficiency will be only 0.5 bps/Hz.

equivalent to 4-PSK (q = 2), has a spectrum efficiency of 2 bits/sec/Hz, and 64-QAM (q = 6) transmits 6 bits/sec/Hz.

Energy per bit

The common measure of digital transmission accuracy is the *bit error rate* (BER), which is the probability of an error occurring, expressed as a negative power of ten. For instance, a BER of 1×10^{-9} means that the probability of making a one-bit error is one part in a billion. At the heart of digital transmission systems is the use of decision circuits that determine the value of a received symbol. In a sense, the BER measures how easy it is for the decision circuit to distinguish between states.

The state diagrams give us insight into this connection. Plots of the state diagram can be generated on a vector oscilloscope, as shown in Figure 3.11 for a QPSK data stream. The scope portrays an accumulation of points corresponding to signal amplitude vs. phase angle. Ideally, we should see four sharp dots; in fact we see blurring of the state loci. This is an indication of noise in the transmission channel, which displaces the signal points, as can be seen by comparing Figure 3.11a with 3.11b. It is easy to understand that the decision circuit can be confused by noise pulses that move one state close to another. The way to minimize that confusion is to increase the signal power without increasing the noise power. In the diagram, this is equivalent to moving the constellation points radially outward, but not increasing the area of blurring.

The key concept here is that, if noise is kept constant, more energy per bit will improve the BER by moving the states farther apart. By extending this idea, we can see a way to compare the performance of the different modulation schemes. For given signal and noise powers, it is clear that the modulation method that puts the greatest distance between its constellation points will have the best BER. This can be seen by comparing QPSK (Figure 3.9) with 16-QAM (Figure 3.10). Considerably more signal amplitude is required in 16-QAM if the states are to be separated as widely as for QPSK. The requirement for more signal power is the price paid for the higher bandwidth efficiency of 16-QAM.

We have established the conceptual basis that connects desired BER with ambient noise level and with required energy per bit. Conventionally, error proba-

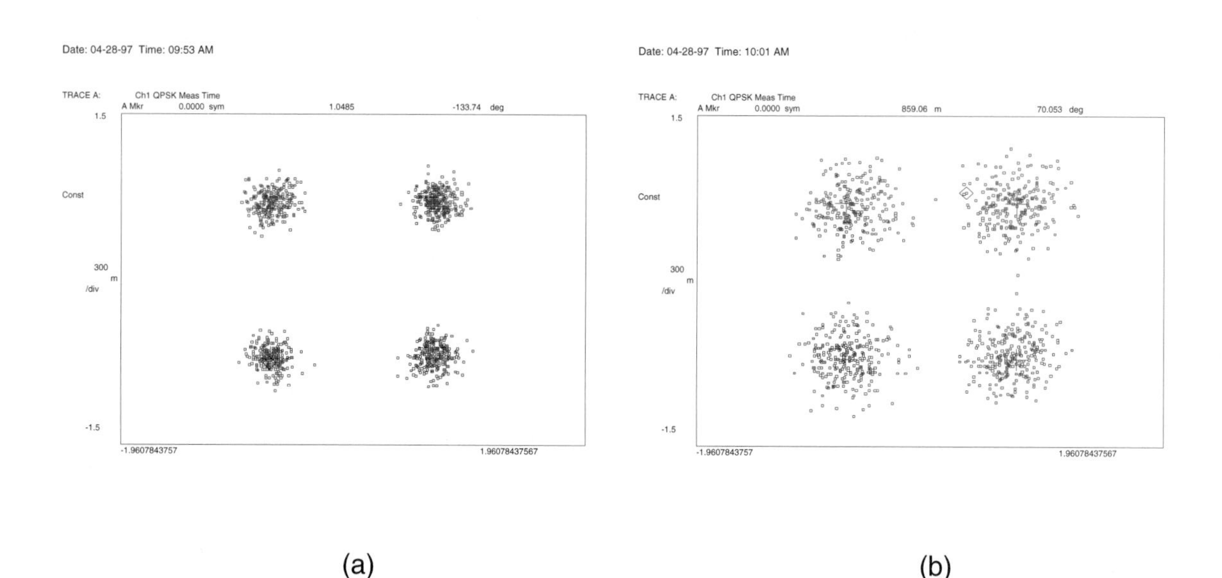

Figure 3.11 Actual state diagrams for QPSK transmissions: (a) C/N = 28 dB, BER = 1×10^{-9} and (b) C/N = 11 dB, BER = 6×10^{-4}

bility is expressed as a rather complicated function of a quantity denoted E_b/N_0, where E_b is the signal energy per bit, and N_0 is the noise power density (noise power in a 1 Hz bandwidth). That relationship can be used to calculate what value of E_b/N_0 is needed to achieve a desired BER. Once E_b/N_0 is known, the carrier-to-noise ratio (C/N) required for that BER can be calculated by the simple relationship

$$C/N = (E_b/N_0)\ (R/B),$$

where B is the noise bandwidth associated with the filter in the detector, and R is the bit rate (bits/sec). Figure 3.12 is a set of so called *waterfall curves*, which show the C/N that is required to achieve given BER performance levels for several different modulation schemes. A second comparison is given in Figure 3.13, which shows both the spectral efficiency and C/N requirement for a variety of modulations at a BER of 10^{-6}. In Figure 3.13 we have noted the fundamental limits derived from Shannon's work in information theory.[11] General statements about the performance of different modulation types is given in Table 3-1.

11. C.E. Shannon, "Communication in the Presence of Noise," *Proceedings of the IRE* 37 (1949), 10–21.

Table 3-1 Performance comparisons

	Bandwidth efficiency	Required C/N	Interference rejection	Complexity
FSK	Low	Low	Medium	Low
PSK	Medium	Medium	Medium	Low
ASK	Medium	High	Medium	Low
M-FSK	Low	Low	Medium	Medium
M-PSK	High	High	Low	Medium
M-QAM	High	Medium	Low	High

Helpful rule-of-thumb:

Shannon predicts that the maximum data-carrying capacity C of a communication channel (in bits/sec per Hz) is:

$$C = B \text{ x } \log_2 (1 + CNR),$$

where B is the bandwidth (in Hz) and CNR is the carrier-to-noise ratio (expressed as a ratio, not in dB). In practical modulation systems, the capacity is about half the maximum attainable, as can be seen by inspection of Figure 3.13. This allows us to estimate the data capacity of a practical link—assuming that we deploy the most efficient modulation scheme that is permitted by the C/N.

If the CNR is 100 or greater (that is, the C/N is 20 dB, or greater), we can simplify the expression by ignoring the 1. When we put in the factor of 1/2 for practicality (noted above) and convert to base-10 logarithms, we get

$$C_{practical} \approx (B/2) \text{ x } [\log_{10}(CNR) \div \log_{10}2] \approx (B/6) \text{ x } C/N,$$

where now we have expressed the carrier-to-noise in the conventional decibels. This means, for example, that a 6 MHz band should be expected to carry the same number of Mbit/sec as its C/N. The rule works because higher-efficiency modulation schemes can be used with higher C/Ns. Alternatively, the expected bits/sec per Hz approximately equals the C/N divided by 6.

Techniques for Dealing with Noisy Channels

A number of very useful approaches are being implemented to increase the chances of communicating successfully through real-world channels. The next

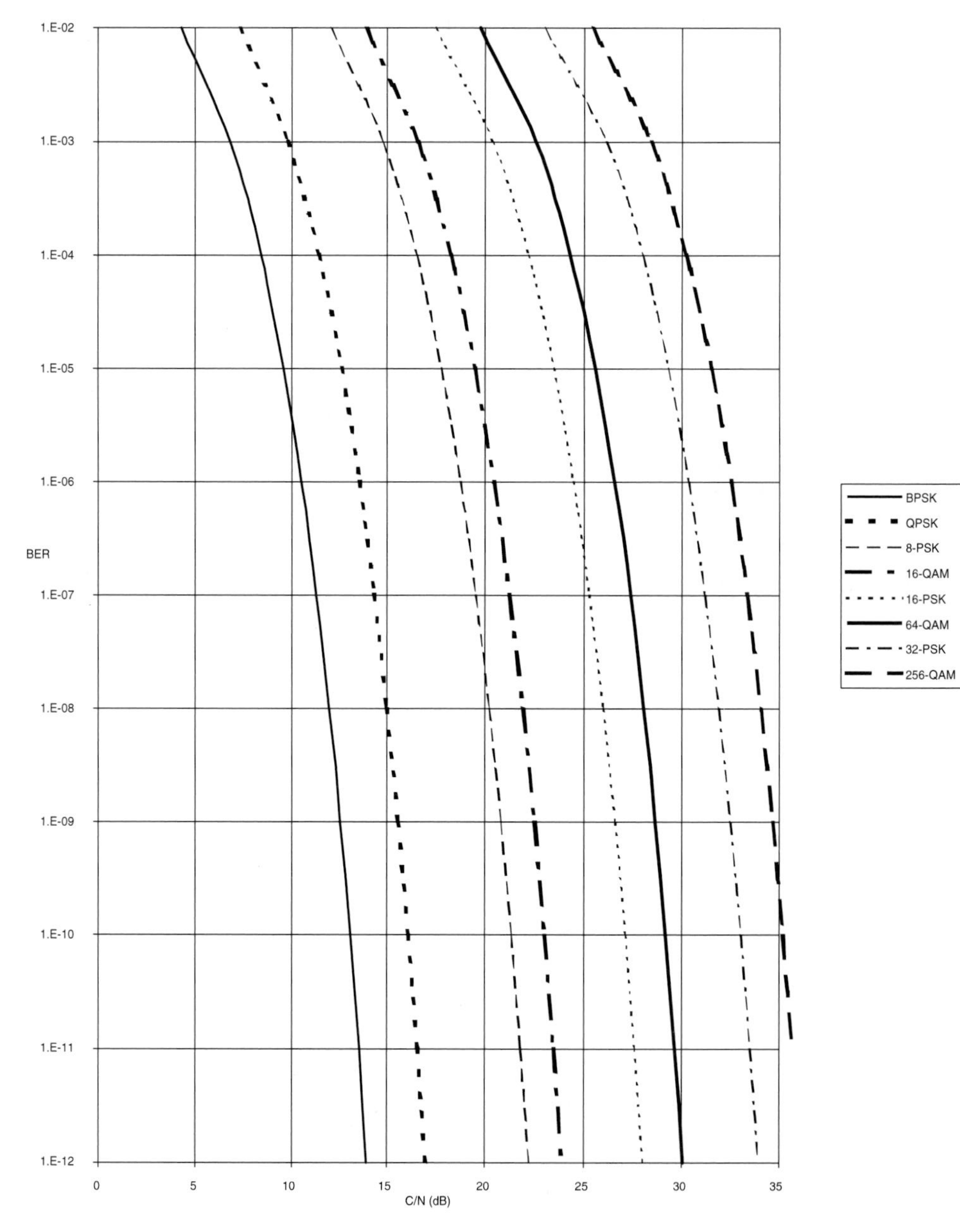

Figure 3.12 Waterfall curves

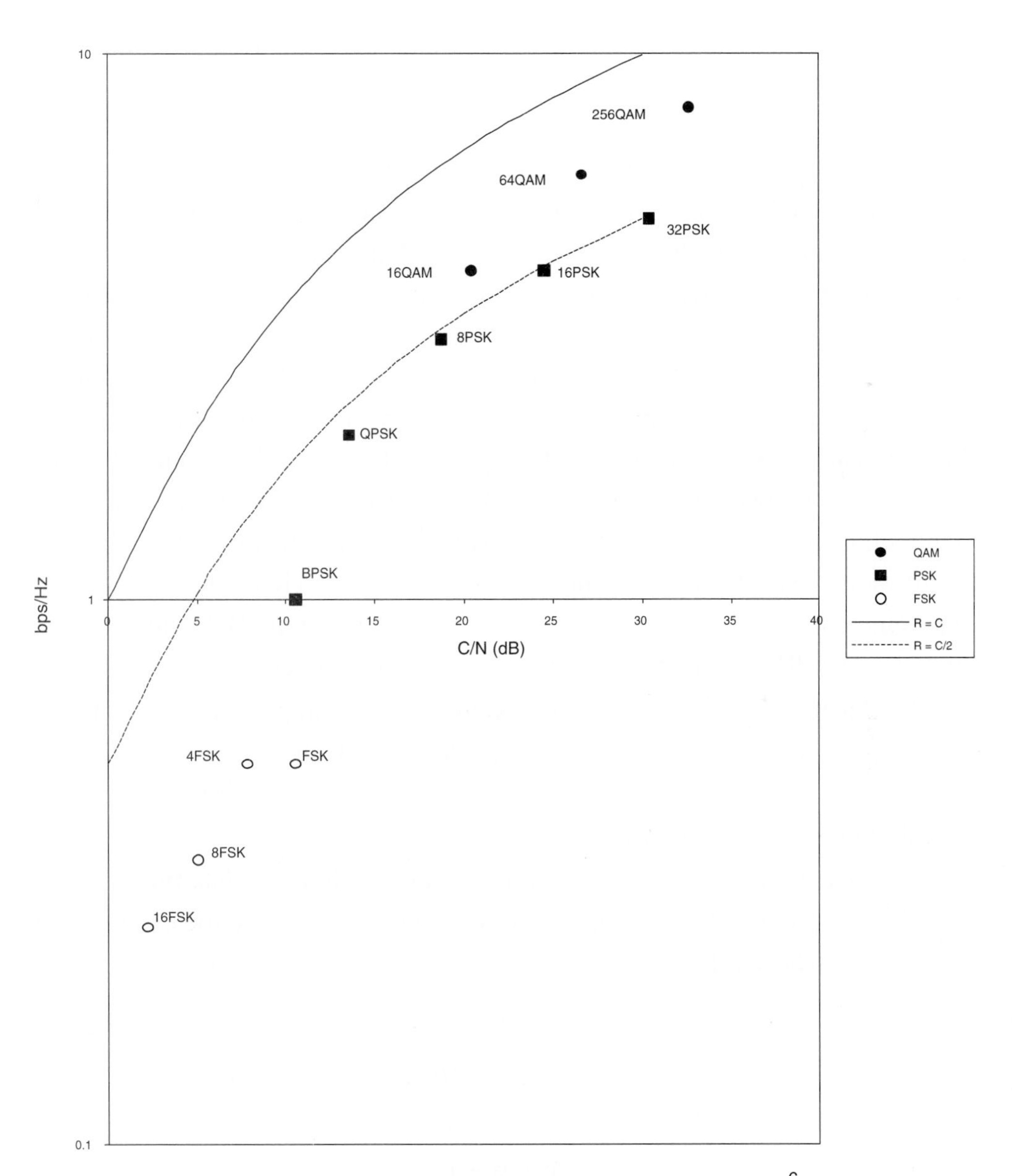

Figure 3.13 Bandwidth efficiency vs. C/N requirement for a BER of 1×10^{-6}. The R = C curve shows the maximum possible rate predicted by Shannon.

three sections examine methods for overcoming noise, interference, and reflections, respectively.

Forward error correction

A considerable body of technology has been developed to detect and to correct transmission errors due to noisy channels. Many of these techniques are used in return path communication systems. The basic concept is to add redundant bits to messages and to use these extra bits, first, to determine whether or not the transmission is accurate and, second, to correct the errors when necessary. In terms of our previous discussion, this technology allows a more marginal C/N (higher chance of error) at some cost in spectral efficiency, since some of the bits transmitted are "overhead" bits, rather than user data. The objective is to find highly efficient *forward error correction* (FEC) schemes that minimize the number of redundant bits. The amount by which the FEC scheme reduces the C/N requirement is called the *coding gain*, which is expressed in dB. Coding gains on the order of 5 dB are routinely achieved for megabit-per-second rates.

FEC technology is a fascinating, extensive, and complex subject. Since our objective is merely to introduce the subject, we will offer an example of one simple FEC scheme in hopes of convincing the reader that FEC is not magic, even though it may seem uncanny.

We give an instance of a cyclic block code[12] with three data bits and three redundant or *parity* bits. This and most other binary codes use the *exclusive OR* operator $\oplus$, whose rules are

$$0 \oplus 0 = 0$$

$$1 \oplus 1 = 0$$

$$0 \oplus 1 = 1$$

$$1 \oplus 0 = 1.$$

If a, b, and c are the three data bits, then (in our example scheme) the parity bits d, e, and f are generated by the rules

12. V. K. Bhargava, "Forward Error Correction Schemes for Digital Communications," *IEEE Communications Magazine* (Jan 1983), 11–19.

$$a \oplus c = d$$

$$a \oplus b = e$$

$$b \oplus c = f.$$

A coder can be constructed to form the data bits and the parity bits into a 6-bit *codeword*, using one 3-bit shift register and two modulo-2 adders (Figure 2 of Reference 12). There will be $2^3 = 8$ "legal" words: 000000, 001101, 010011, 011110, 100110, 101011, 110101 and 111000. We can now see where the coding gain comes from. We are using only 8 states out of the total of 26 = 64 states that are possible in 6-bit symbols. This effectively increases the space between states. On the other hand, the added parity bits mean that our effective data rate has been halved.

We will illustrate a simple error detection and correction scheme: table look-up decoding. When the information is received, the three parity bits are calculated (from the three data bits). The calculated parity bits are then exclusive-OR'd with the received parity bits. The result is called a *syndrome* since, in a sense, it may characterize a "disease." The syndrome is used to address a ROM-based table of correction words (Table 3-2). When the appropriate correction word is exclusive OR'd to the received word, the correct transmission is re-created.

Table 3-2 Correction table

Syndrome	Correction
000	000000
001	000001
010	000010
100	000100
101	001000
011	010000
110	100000
111	100001

For example, if 001101 is sent, but 011101 is received, the calculated parity bits are 110, and the syndrome is 110 ⊕ 101 = 011. The table gives a correction word of 010000. When this is exclusive-OR'd to the received number, we get 010000 ⊕ 011101 = 001101, which is the original transmission.

The scheme illustrated in this example is intended to correct only single-bit errors. A method as slow as table look-up works well in our example because the number of different 6-bit patterns with one or fewer errors is so limited. Other, more complex coding schemes are capable of correcting multi-bit errors. As the complexity goes up, the table look-up correction method used in our example becomes less and less practical. A number of popular FEC methods, such as *Reed-Solomon coding* and *trellis coding*, are much more efficient.

Interleaving

Error correction works best when you can be sure that any symbol will have no more than one or two errored bits. Unfortunately, a spike of noise is capable of obliterating a series of consecutive bits. A method for reducing the effect of such spikes is called *interleaving*. It mitigates the effects of bursts of interference and noise by distributing the bits from one symbol over a longer period of time, which increases the chances for either the data bits or the parity bits getting through uncorrupted. This gives the message a higher probability of being correctly reconstituted. Specifically, each symbol is read into a matrix as one of N rows prior to transmission, but the *columns* of the matrix are read out for transmission (Figure 3.14). The number of rows is an indicator of how much spreading is being done and is referred to as the *interleaver depth*. The downside of this idea is that some delay needs to be built into the transmission in order to create the interleaving matrix and to reconstruct the symbols at the detector. In many applications, such as data communications, moderate delays are acceptable. On the other hand, some applications—such as interactive video games, where delays of as little as 10 msec are objectionable—are very delay-sensitive and require very short interleaver depths.

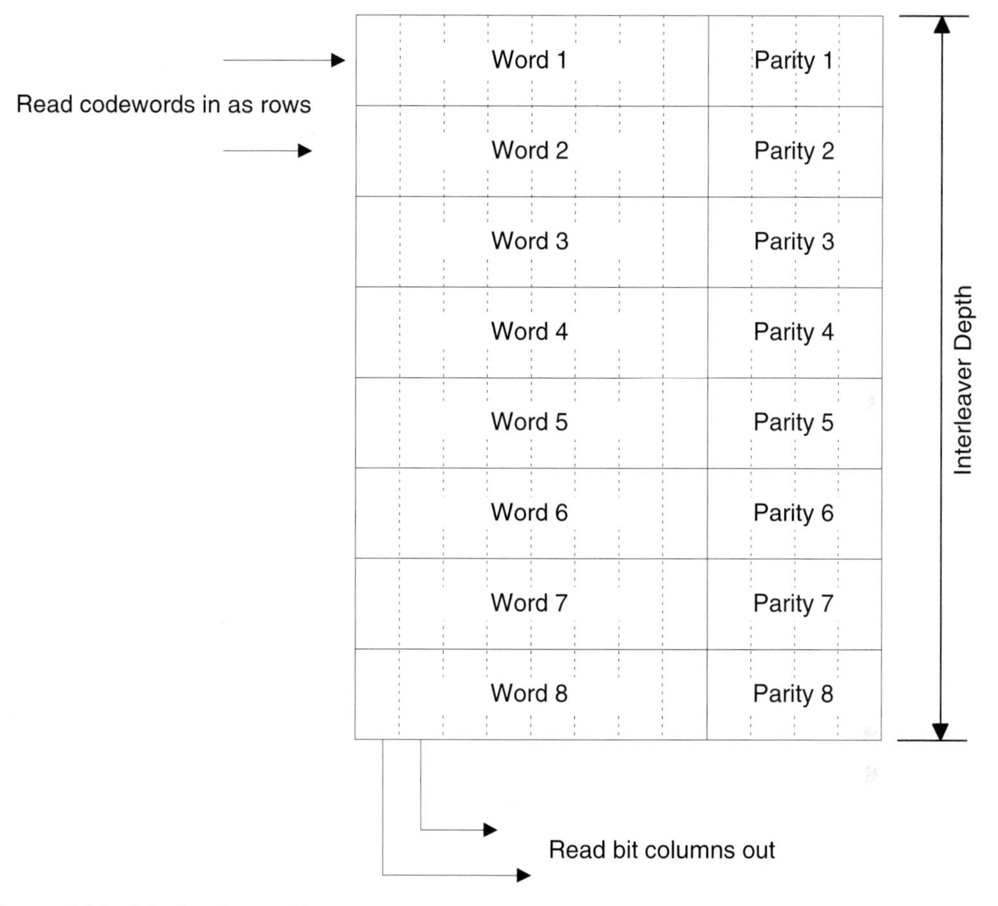

Figure 3.14 Interleaving matrix

Equalization

Imagine the challenge of accurately detecting, for example, a 16-QAM signal—which depends on both amplitude and phase fidelity—after it has passed through a transmission path that has a number of reflection points. The detector sees not just the desired signal but also a series of attenuated and delayed images. No matter how well-maintained the cable plant is, the potential for reflection problems will always exist in subscribers' in-home cabling. Fortunately, there is a signal processing technique—equalization—that can minimize the effects of transmission-path reflections. The underlying technology dates back to work in radar systems, but integrated circuits and digital signal-processing techniques have made it readily applicable to consumer-level products.

The concept is straightforward. Since our need is to remove attenuated and delayed copies of our desired signal, we can pass our received composite signal through a device that subtracts attenuated and delayed signals. This miraculous-sounding device—called a tapped delay line—is actually quite simple. As diagrammed in Figure 3.15, the output is the sum of the received signal f(t) plus a number of copies of f(t), each delayed by one bit period τ and attenuated by a set of coefficients α_i.

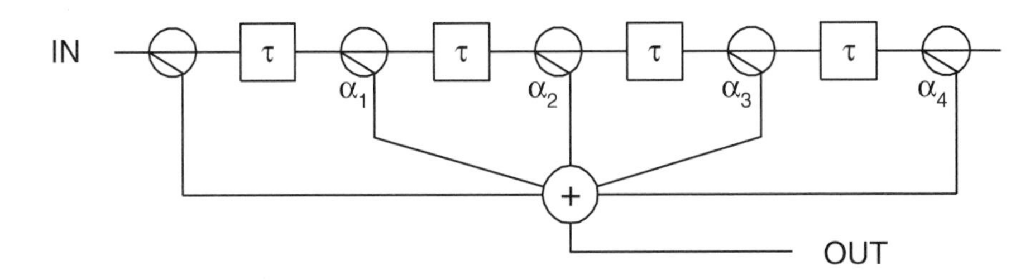

Figure 3.15 Block diagram of a 4-tap equalizer

The hard part, of course, is to determine the correct values for these coefficients. One method is to transmit a *training sequence* consisting of a known series of bits and to use well-established routines for converging the coefficients.

Words of Caution

In this chapter we have discussed the performance requirements for achieving a given error rate in an ideal communication channel that contains *additive white gaussian noise* (AWGN). Two important cautions need to be raised at this point. The first is that the channel may be degraded by non-gaussian noise and interference, as we will discuss in the next chapter. The various modulation types will have differing robustness in such cases. Second, the operator should expect the performance of commercial hardware systems to be lower than that predicted by Figures 3.12 and 3.13, which in turn will require better noise performance in the channel. There are many good reasons for this, such as the need for the application receiver to filter and amplify the received RF signal. The loss through the filter and the noise figure of the pre-amp (just two components of the *implementation loss* of the system) have the effect of raising the noise figure of the detection circuit, which reduces the C/N prior to the detector. The C/N decrease may be as much as 10 dB, which is the same as raising the channel C/N requirement by 10 dB.

Summing-Up...

- A multitude of high-bit-rate modulation schemes are being deployed successfully.
- Modulations with higher bandwidth efficiency require higher C/N in the channel.
- The various modulations have different sensitivities to different types of channel impairments.
- Forward error correction techniques reduce the C/N requirement, at relatively little cost in data rate.
- Equalization and interleaving can increase margins for reflections and noise bursts, respectively.

CHAPTER 4

Sources of Noise and Interference

The previous chapter explained how the designer of digital terminal equipment determines the carrier-to-noise performance requirement for transmission over the proposed communication path. In a sense, this chapter discusses the phenomena that make it difficult for the cable TV operator to provide that level of C/N for transmission over the return path. Rest assured, however, that the remainder of the book will deal with the key design and operation techniques to ensure that these transmissions will arrive reliably and accurately.

So let's "boldly go forth" to meet our enemies—noise and interference.

Return Noise Funneling

The first aspect of our problem is that the basic architecture of the return path system makes the job more difficult. To fully appreciate this, let's compare the forward path with the return. In the forward path, there is essentially only one location where signals enter the distribution plant—the headend—and the operator has tight control of the signal at that point. From the headend to the subscriber, the signal fans out in a point-to-multipoint array. It's just the opposite for the return path: signals can enter the return plant from every home that is attached to the plant, and all of those signals fan-in (combine) as they travel toward the headend. This is why some people refer to the return plant as a "noise funnel." Fortunately, the hybrid fiber/coax (HFC) system design gives the operator some level of control by allowing the system to be segmented into individual fiber nodes (Figure 4.1), unlike the earlier tree-and-branch architecture, which hardwired essentially all the subscribers together.

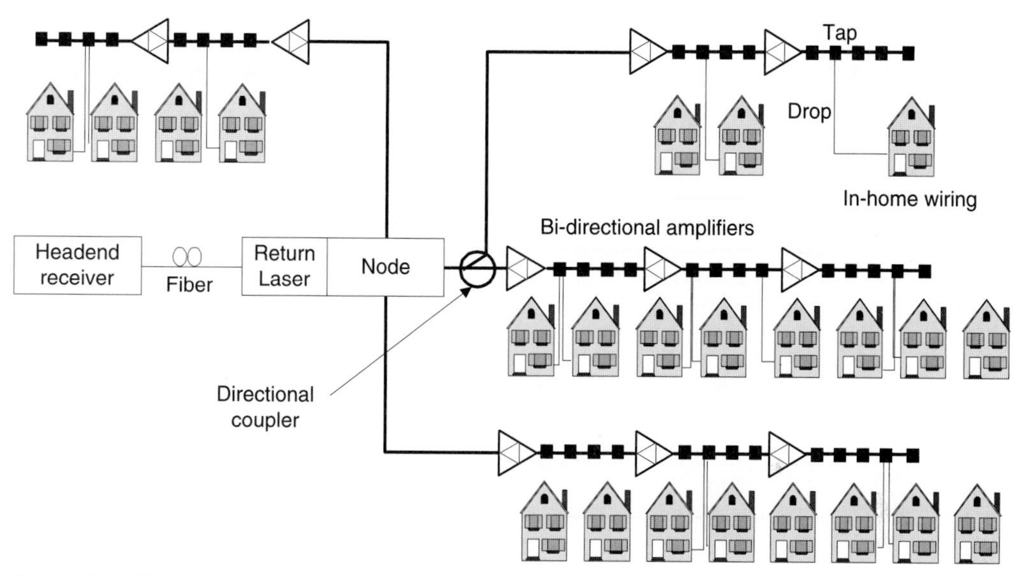

Figure 4.1 Return distribution through HFC node

There are three primary sources of noise in the return path—thermal, fiberoptic link, and ingress. As shown in Figure 4.1, thermal noise is generated in each of the active components (amplifiers and receivers); fiberoptic link noise is generated in the return laser, fiber, and headend receiver; and ingress comes from the surrounding environment, but generally enters the system through the in-home wiring and the drop cabling. We will examine each noise source in detail.

Thermal Noise

Amplifiers generate noise due to thermal fluctuations of the electron density. This is usually characterized by the *noise figure* of the amplifier, NF. If a noise-free carrier signal is transmitted through an amplifier, then the carrier-to-noise ratio (C/N) is given by the formula

$$C/N_{amp} = \text{Input} - \text{NF} - (\text{Noise floor}),$$

where the input is in dBmV. The *noise floor* is the ideal thermal noise power in the bandwidth of interest (generally referred to as the *noise bandwidth*) and expressed in dBmV. The noise floor can be calculated for any noise bandwidth by the formula

$$\text{Noise floor} = 10*\log(\text{Noise bandwidth, in Hz}) - 125.2 \text{ dBmV},$$

where –125.2 dBmV is the thermal noise in a 1 Hz bandwidth for a 75-ohm system.[1]

Table 4-1 Noise floor

Noise bandwidth	Noise floor (dBmV)
100 kHz	–75.2
200	–72.2
300	–70.4
400	–69.2
500	–68.2
600	–67.4
700	–66.7
800	–66.2
900	–65.7
1 MHz	–65.2
2	–62.2
3	–60.4
4	–59.2
5	–58.2
6	–57.4

1. Strictly speaking, this is the thermal noise at room temperature. The value changes by ±0.5 dB for a rise or fall of 40°C, but this is not within the measurement accuracy of field instruments.

For the forward band, we are used to seeing the noise floor given as –59.2 dBmV (in NTSC systems) or as –58.2 dBmV or +1.8 dBμV (in PAL systems). This is because the forward video signals have a standardized noise bandwidth—4.0 MHz for NTSC, and 5.0 MHz for PAL—which simplifies the formula. As we will see, it is not as easy for the return path, where the upstream signals may have bandwidths varying from 100 kHz to 6 MHz. As an aid to the reader, we have listed in Table 4-1 the noise floor for several noise bandwidths. The noise floor numbers in Table 4-1 can be plugged right into the first formula (keeping in mind the double minus signs).

The noise figure of the integrated circuit used in return amplifiers is on the order of 5 dB. In the multi-port amplifier configurations commonly used for nodes and distribution amplifiers (DA), the return signals from several ports are combined prior to amplification. Each of these internal combines reduces the effective input to the amplifier IC by approximately 3.5 dB. In addition there is approximately 2 dB of input loss due to the diplexer and test point. Accordingly, the *effective* NF of a 4-output amplifier (or node) is about 14 dB, a 2-output (distribution amplifier) about 11 dB, and a 1-output (line extender) about 7 dB. (When doing an actual design, you will want to use the numbers supplied by the manufacturer of the specific equipment.)

Since noise sources are uncorrelated,[2] noise adds on a power basis (that is, as 10*log), so two C/N ratios, $(C/N)_1$ and $(C/N)_2$, can be combined as

$$(C/N)_{total} = -10*\log(10^{-(C/N)_1/10} + 10^{-(C/N)_2/10}) \quad .$$

This formula can be rewritten by factoring out $(C/N)_1$:

$$(C/N)_{total} = (C/N_1) - 10*\log(1 + 10^{-(C/N)_2/10 + (C/N)_1/10}) \quad .$$

The complicated second part of this formula can be calculated and tabulated as a function of the difference between the two C/Ns (Appendix D, Table D-2).

2. It is possible to have correlated noise, when a single interference source is picked up at several points in the system, but this is not the general case. For further discussion, see C.A.Eldering, N. Himayat, and F.M. Gardner, "CATV Return Path Characterization for Reliable Communications," *IEEE Communications Magazine* (August 1995), 62–69.

Amplifiers often appear in cascade. As a return signal travels up the cascade, noise will be added by each amplifier. The combined C/N of n similar amplifiers in series (each with a C/N ratio of $(C/N)_1$) is

$$(C/N)_{total} = (C/N)_1 - 10*\log n \quad .$$

The C/N degrades by 3 dB for each doubling of the cascade.

Example

What is the C/N for an RF return consisting of a node and four amplifier branches, each made up of two DAs and two LEs? The input signal to each amplifier (for the service of interest) is 5 dBmV, with a 200 kHz noise bandwidth.

The procedure is to determine the C/N for each branch, combine the branches, and then add the node. Using the amplifier NFs given above and the 200 kHz noise floor from Table 4-1, the C/N for one branch is calculated as follows:

$C/N_{DA} = 5 - 11 - (-72.2) = 66.2$ and two DAs combine to $66.2 - 10*\log 2 = 63.2$.

$C/N_{LE} = 5 - 7 + 72.2 = 70.2$ and two LEs combine to $70.2 - 3 = 67.2$.

Using Table D-2 in Appendix D, the carrier to noise for the branch is $C/N_{branch} = 63.2 - 1.46 \approx 61.7$.

The four equal branches combine to $61.7 - 10*\log(4) = 55.7$.

The C/N due to the return RF portion of the node is $5 - 14 + 72.2 = 63.2$.

Combining the branches with the node (again using Table D-2), the total C/N for the RF distribution is $55.7 - 0.71 \approx 55.0$ dB.

Fiberoptic Link Noise

In the node, the RF return signal is applied to a laser, thereby amplitude modulating the optical output. The signal passes through kilometers of optical fiber on its way to a photoreceiver in the headend. In the process, several noise sources come into play, which we will lump into one. Chapter Nine discusses fiberoptic links in detail and gives expressions for the individual components, but the system designer really needs only one set of C/N numbers—supplied by the equipment manufacturer—to determine the effect of the fiber link on overall performance. Figure 4.2 shows a representative curve for the C/N performance of a return fiber link as a function of the link length. Note that the link length is given in dB of optical attenuation at 1310 nm optical wavelength, which equates to approximately 0.4 dB/km (2.5 km/dB).

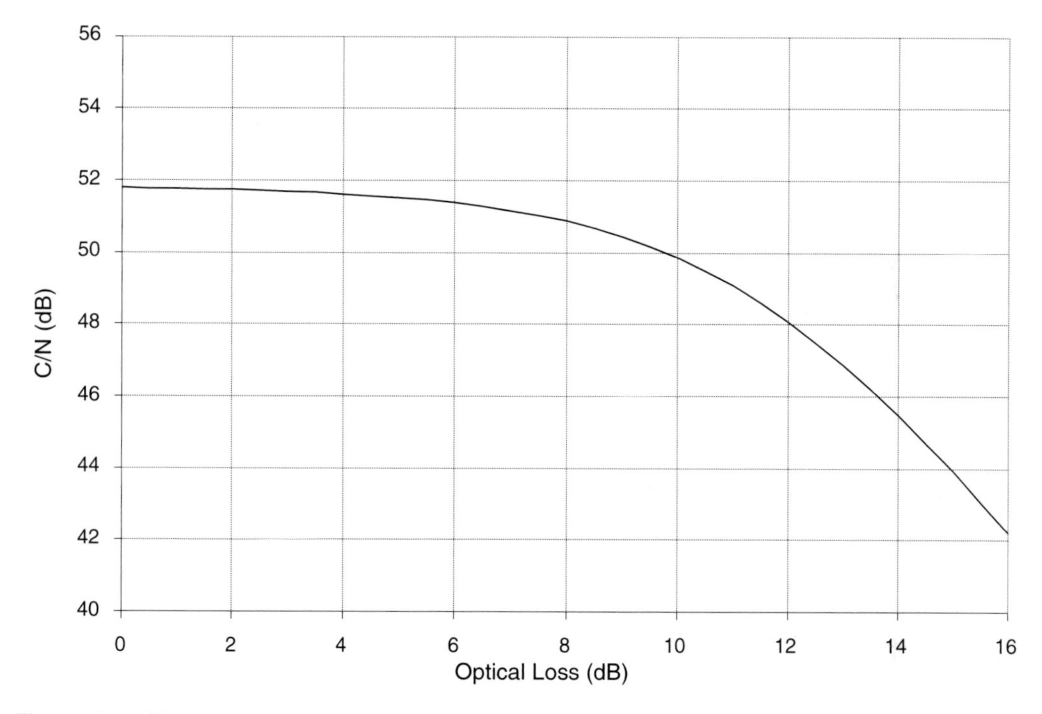

Figure 4.2 Fiberoptic link carrier-to-noise (Fabry-Perot laser, 4 MHz noise bandwidth)

The C/N performance of a fiber link is shown in Figure 4.2. It is unfortunate that the figure—which presents the information in a form that has been fairly typical—is misleading. In Chapter Nine we will discuss the fact that there is a limit on the amount of RF power that can be applied to a laser. In Chapter Ten we will explain methods for allocating that limited power among various applications. Figure 4.2 shows the C/N that would be achieved if all of the power were allocated to a single signal with bandwidth of 4 MHz, while in a more realistic situation the power would be shared equally across the full 5 to 40 MHz return band. This means that the C/N ratios in the figure need to be corrected by a factor of 10*log(4 MHz/35 MHz), as will be explained in the later chapters. That reduces the C/N at a 5 dB fiber loss from 51.5 dB to (51.5 – 9.4)

= 42.1 dB. The method that will be described in Chapter Ten provides the same fiberoptic link C/N for all signals, independent of their bandwidths.

This means that the fiber link causes a major degradation in the overall C/N. In our example, we had calculated an aggregate C/N of 55 for the RF distribution from the node; the fiberoptic link alone has a C/N of 42 for typical link losses. We should not lose sight of what that fiberoptic link is accomplishing, of course, which is to span 16 km (thus replacing 32 amplifiers). Nonetheless, the important point to take away from this discussion is that as far as distribution *equipment*, the C/N is dominated by the fiberoptic link. In our example, the composite C/N at the headend would be

$$C/N\text{total} = \{\text{Sum of } C/N_{\text{fiber}} = 42.1 \text{ and } C/N_{\text{RF}} = 55.0\} = 42.1 - 0.26 = 41.8.$$

Equipment tips

Two caveats are required regarding vendor information about laser performance. First, since one source of C/N degradation is the interaction of the laser with the fiber, it is important to ensure that the C/N curves show performance measured with actual fiber, rather than with an optical attenuator. Second, as a general rule the noise contribution due to the headend receiver will be about the same for all vendor products, which should make it possible to mix and match between node and receiver vendors. Make sure, however, that the receiver noise is included in the C/N curve provided by the transmitter vendor.

Ingress

If our exploration of the sources of C/N degradation were over at this point, we'd be in great shape because, as we saw from the previous chapter, a 42 dB carrier-to-noise is more than sufficient for any conceivable digital transmission. But we are not done; we are about to begin grappling with the nastiest part of our problem: *ingress*.

Sources of ingress

The cable plant exists in a sea of background radiation that includes signals in the return RF frequencies. There are both discrete and diffuse interferers. The discrete sources for North America, such as *citizen's band* (CB) and *shortwave* radios (both broadcast and amateur), are listed in Table 4-2.[3] Amateur (ham) radio is a concern because its relatively strong signals can occur at random times and at varying frequencies throughout the HF spectrum, which overlaps much of the return band below 30 MHz (Table 4-3). Since essentially all of these signals are transmitted by bouncing a radio wave off the ionosphere, they are subject to the conditions of that ionized layer at the particular time—whether it is absorbing or reflecting the desired transmission frequency. The band of frequencies that will be reflected efficiently at any moment depends on such unpredictable phenomena as the intensity of ionizing radiation from the sun.

Table 4-2 Discrete interference sources in the return band (North America)

Shortwave broadcast	Voice of America
	Radio Moscow
	Radio Havana
	Radio Marti
Other over-the-air	Amateur radio
	Citizen's band
	Paging transmitters
In-home	Home intercom using AC power wiring

The intensity of this radiation changes daily. As a result, for instance, the bands below 9 MHz are generally not usable for amateur radio during the daylight hours. Whether or not a band can be used depends also on the longer-period variations of the sun, specifically its rotational period of 27 days and its *sunspot cycle* of approximately 11 years, which is shown in Figure 4.3.[4] The sunspot cycle is predicted to peak in the year 2000, which is likely to result in

3. Shortwave broadcasts tend not to be on constant frequencies. Schedules of time and frequency for various languages of programming to be sent out by broadcasters from around the world are published regularly. A list of Internet sites for programming information is maintained by Radio Netherlands at http://www.rnw.nl/en/pub/hitlist.html.

4. The figure shows *sunspot number*, which is a weighted average of measurements of sunspot quantities as recorded by a number of observatories throughout the world. Sunspot numbers have been recorded since the invention of the telescope in 1610. The NGDC website (see Fig 4.3) shows data beginning in 1700. The polarity of the sun's magnetic field reverses at each maximum of the sunspot cycle.

Table 4-3 U.S. amateur bands at return frequencies

Band (meters)	Frequencies (MHz)	Maximum power (watts PEP)
6	50.000 – 54.000	1500
10	28.000 – 29.7000	1500 [a]
12	24.890 – 24.990	1500
15	21.000 – 21.450	1500
17	18.068 – 18.168	1500
20	14.000 – 14.350	1500
30	10.100 – 10.150	200
40	7.000 – 7.300	1500

a. Lower license rating limited to 200 W PEP (peak envelope power)

two problems for the cable return system. First, the overall level of HF radiation on earth will increase, which could cause an increase of diffuse ingress in the lower portion of the return band. Second, as amateur radio propagation in the lower bands deteriorates, ham operators are likely to move to the bands in the 15–30 MHz range (i.e., to shorter wavelengths). This is a double-whammy for the cable system operator since the solar radiation will have direct adverse impact on the lower half of the return spectrum and an indirect one on the upper half. Furthermore, we have no quantitative estimates of these impacts since that would be akin to predicting the weather for the year 2000.

Another important source of diffuse background energy is *impulse noise*, consisting of short (few microsecond) pulses of broadband emissions that are generated by a great variety of common phenomena, including arcing, switching transients, and intermittent grounds. These impulses arise from large-scale power distribution as well as from individual household appliances.

Entry points for ingress

At first thought, it would seem difficult for unwanted signals to get into a communications system that is composed of coaxial cable, well-shielded ampli-

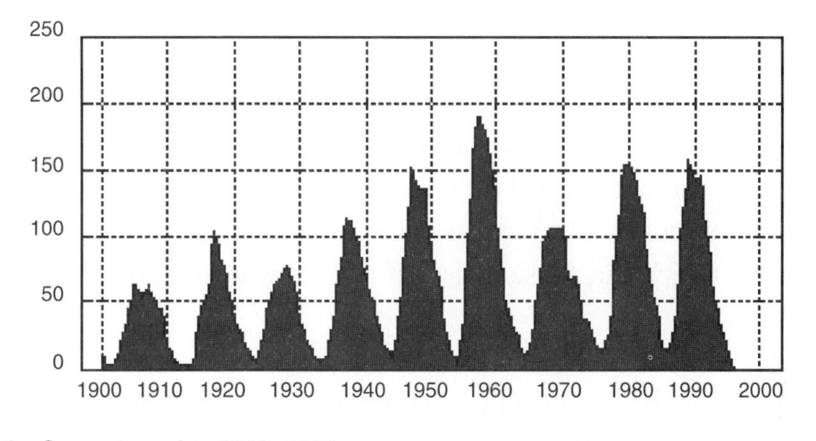

Figure 4.3 Sunspot number 1900–1995

Source: National Geophysical Data Center web site http://www.ngdc.noaa.gov.

fiers, and optical fiber. In the US there are very strict regulations on the amount of RF energy that can be *emitted* from the cable plant, so the distribution equipment and cabling must be maintained to a high degree of RF "tightness."After further thought, however, one realizes that there are several semi-porous portals in the cable system's shield against RF ingress.

The subscriber's home is the most permeable part (Figure 4.4). It is common to hear estimates that 50 percent or more of the total system ingress enters through the home. Ingress starts at the back of the TV set, which is not designed to provide sufficient rejection of ambient signals in the return band frequencies.[5] It also enters through unterminated splitter ports[6] and through poorly connectorized in-house cables. Because most consumer-level products have very poor *return loss,*[7] it is likely that there are multiple reflections of signals within the in-house cabling. TV tuners and set-top boxes generally have adequate return loss only in the channel being tuned, which compounds the problem of reflec-

5. Only recently have TV set manufacturers paid proper attention to *direct pick-up* of signals within the forward band, which is why one can see over-the-air signals ghosting with cable-delivered signals on many sets.

6. Next time you are in a store that sells TV splitters, look for the display of 75-ohm terminators. You may have to look hard.

tions at all other channel frequencies. These reflected signals can be directed up the drop cable, where they commingle with the desired return signals.

Impulsive signals in the home—including TV on-off switch transients[8]—are likely to couple into the coaxial cable shield and go to ground through the in-home cabling.[9] It is not unusual for the cable system to provide the best grounding path in its vicinity. (This occurs on the main distribution portions of the plant, as well, which can cause major difficulties when the local electric utility drops a phase.)

An additional 20 to 30 percent of the ingress enters through the drop system. This happens because (a) the F-connectors can degrade over time if they are improperly installed, (b) the coax shield does not provide 100 percent coverage when new[10] and becomes porous to RF with age due to continual flexing and corrosion damage, and (c) the maintenance crews have a hard time keeping up, since there are so many drops. Figure 4.5 indicates additional ingress and distortion sources.

We should take a moment to entreat the reader not to lose hope. The problems sound daunting, but there *are* solutions. In the next chapter we will discuss methods for controlling ingress.

7. Return loss is the amount by which the reflected signal power is lower than the original signal (see Appendix D). In distribution equipment, return losses of 14–16 dB across the entire cable TV spectrum are the rule. Consumer grade splitters generally have less than 10 dB return loss, which means that more than 10% of the input power gets reflected back at a splitter port.

8. R. Hranac, "Impulse Noise in Two-Way Systems," *Communications Technology* (July 1996), 16–20.

9. Such signals are called *common-mode*, as opposed to the normal signals in coaxial cable (called *differential-mode* signals), which appear only in pairs: the shield current being accompanied by an equal signal in the center conductor that is 180° out-of-phase. Since common-mode signals carry a net current, they can be detected by a sensing loop surrounding the cable. This is a good tool for isolating that class of ingress problem.

10. Tri-shield drop cable is available. A braid coverage of 90% is estimated to give 108 dB of isolation, and 60% gives about 100 dB.

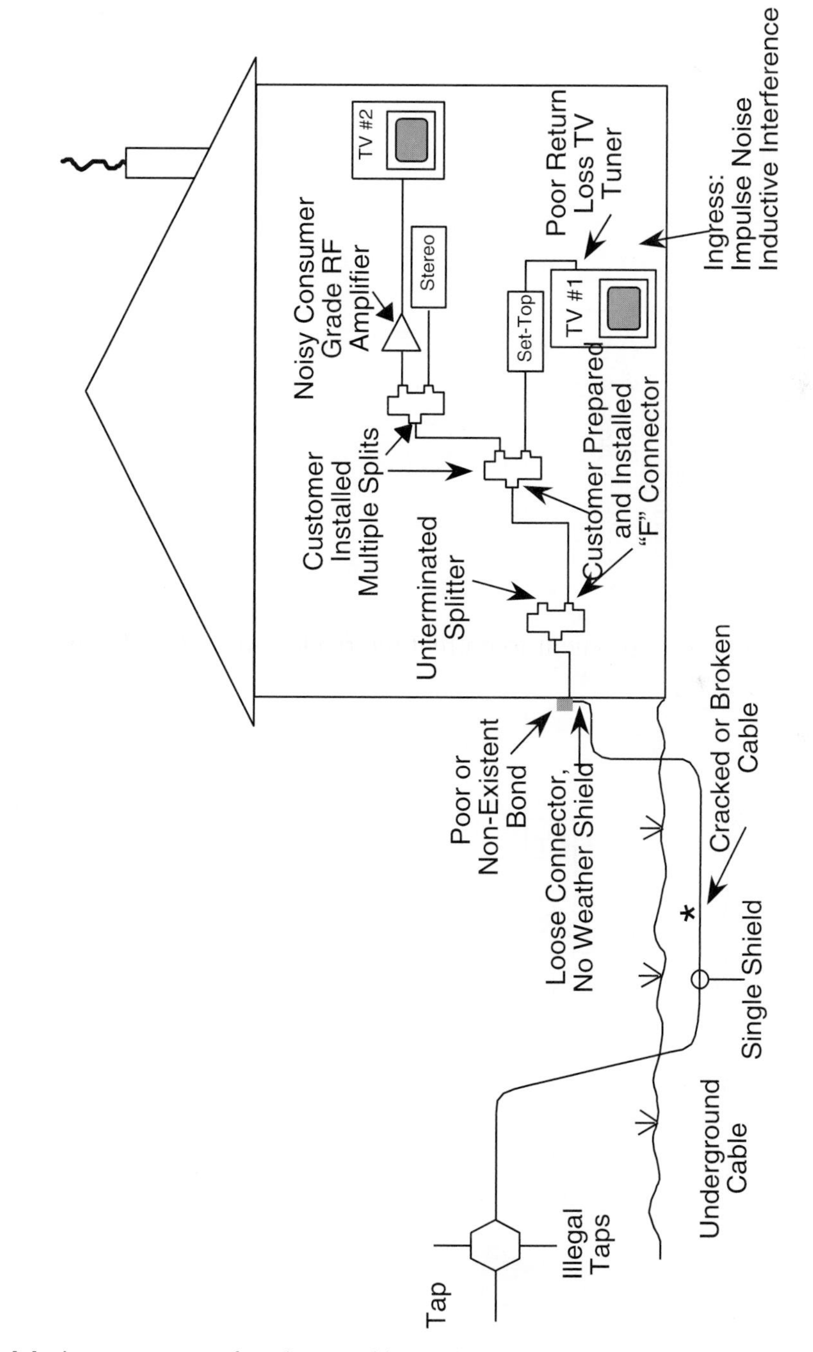

Figure 4.4 Ingress sources from house wiring to the tap

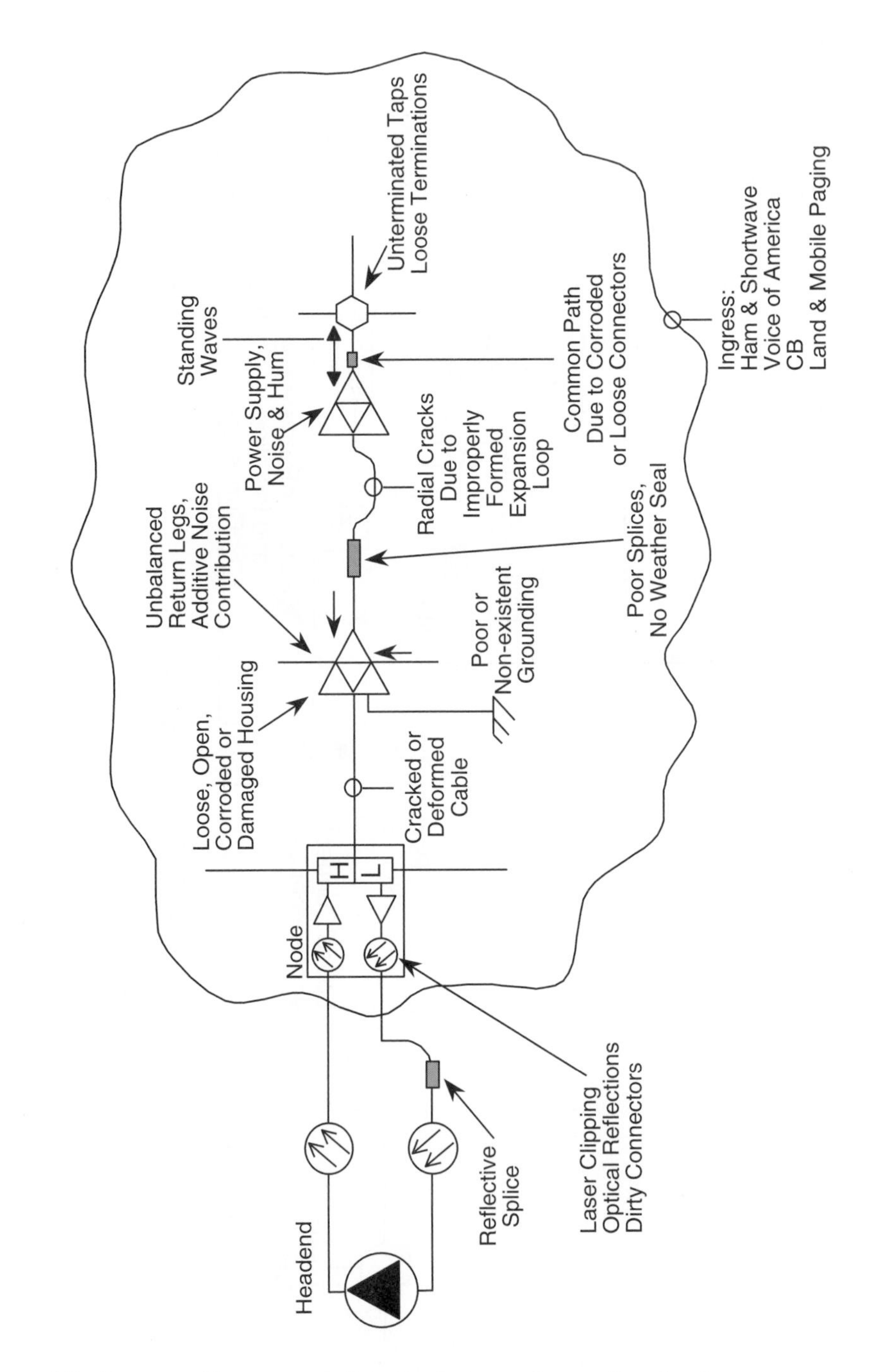

Figure 4.5 Sources of ingress and distortion between tap and headend

Distortion and Clipping

As in the forward path, non-linearities in amplifiers and lasers will cause *intermodulation distortion* of the signals. They will also cause *clipping distortion,*[11] which occurs when the RF signal impressed on a laser is large enough to drive it below threshold (Figure 4.6).

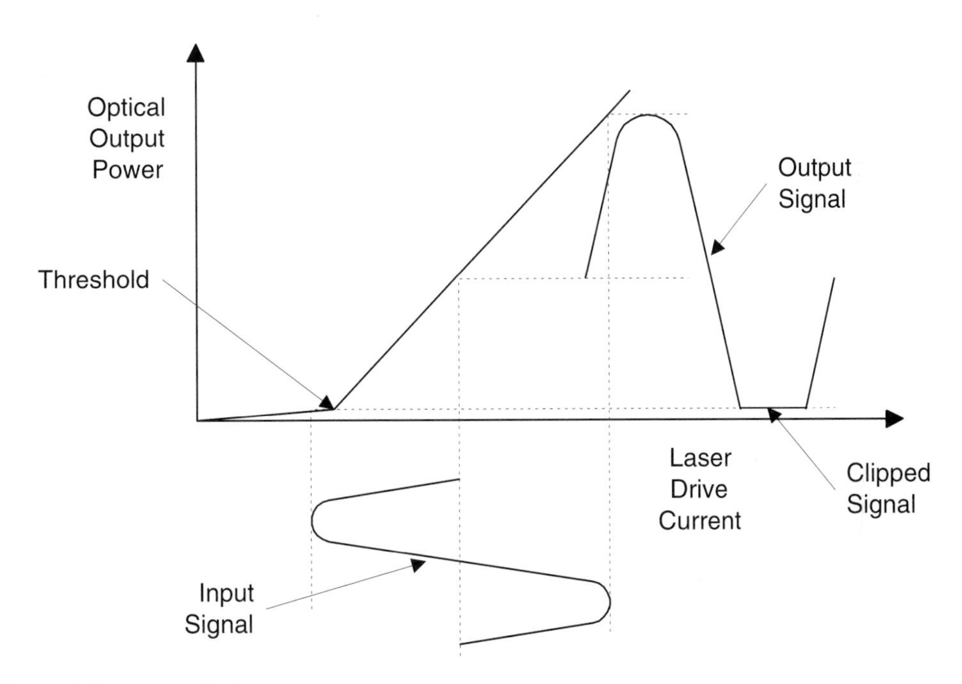

Figure 4.6 Laser clipping

Unlike the forward path, however, the return path distortions do not show up as the typical beat packet. Rather, they appear very much like noise. This is because the digital signals carried on the return path look, themselves, like blocks of noise (Figure 4.7), and when noise-like blocks beat together, they produce intermodulation products that also appear noise-like.[12] This has led people

11. D. Raskin, D. Stoneback, J. Chrostowski, and R. Menna, "Don't Get Clipped on the Information Highway," NCTA Technical Papers, pp. 294–301, 1996.

12. J. Hamilton and D. Stoneback, "The Effect of Digital Carriers on Analog CATV Distribution Systems," NCTA Technical Papers, pp 100-111, 1993.

to use the term *composite intermodulation noise* (CIN) to refer to the distortion produced by digital signals. CIN is analogous to CTB and CSO, which are measures of distortion for discrete tones.

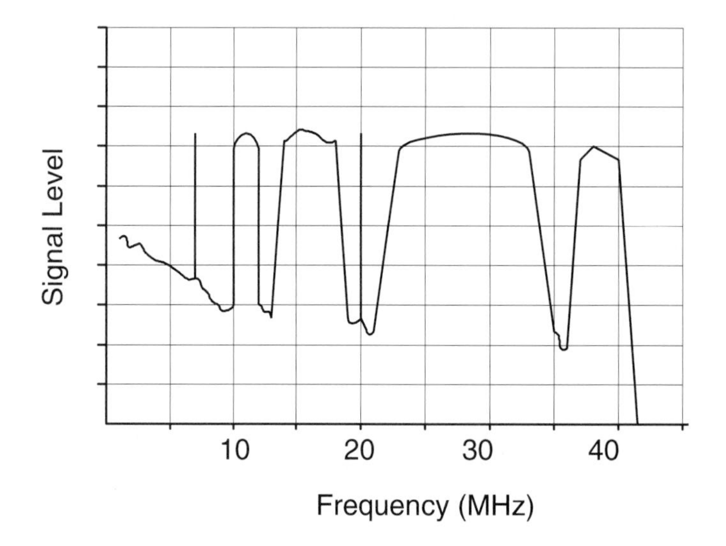

Figure 4.7 Return band digital signals

This can be seen experimentally by passing a 5–40 MHz band of noise (to simulate many data signals) and one actual QPSK signal through a return laser. The 2 MHz-wide QPSK signal (2 Mbps) is placed in a filtered slot in the noise, as shown in Figure 4.8a. When the QPSK signal is turned off, the depth of the empty slot shows the inherent noise of the fiber link (42.2 dB below the signal level). When the level of the signal and noise block is then increased by 10 dB (Figure 4.8b), it appears as though the noise floor in the slot is rising. In actuality, this rise amounts to 24dB and is due to intermodulation distortions in the laser. A further 6 dB signal level increase (Figure 4.8c) causes nearly an additional 20 dB rise in the floor. Note the slot filling appears to be noise-like, and the increase is neither the 2:1 ratio that would be predicted from pure second-order distortion nor the 3:1 expected from pure third-order. The slot is filling, in fact, with a composite of many even- and odd-order intermodulations.

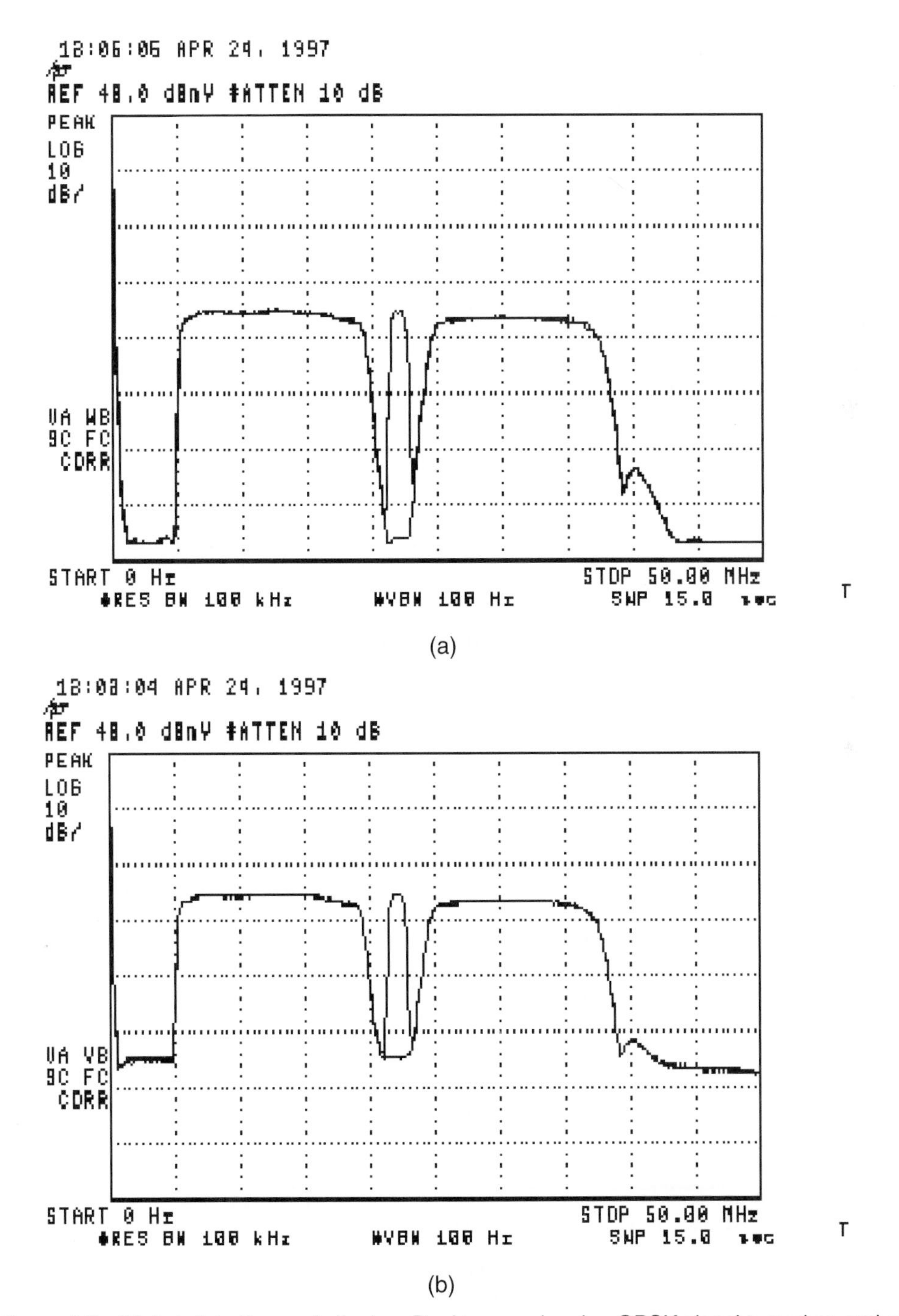

Figure 4.8 Digital distortion and clipping. Dual traces showing QPSK signal turned on and off (a) nominal loading, (b) nominal + 10 dB

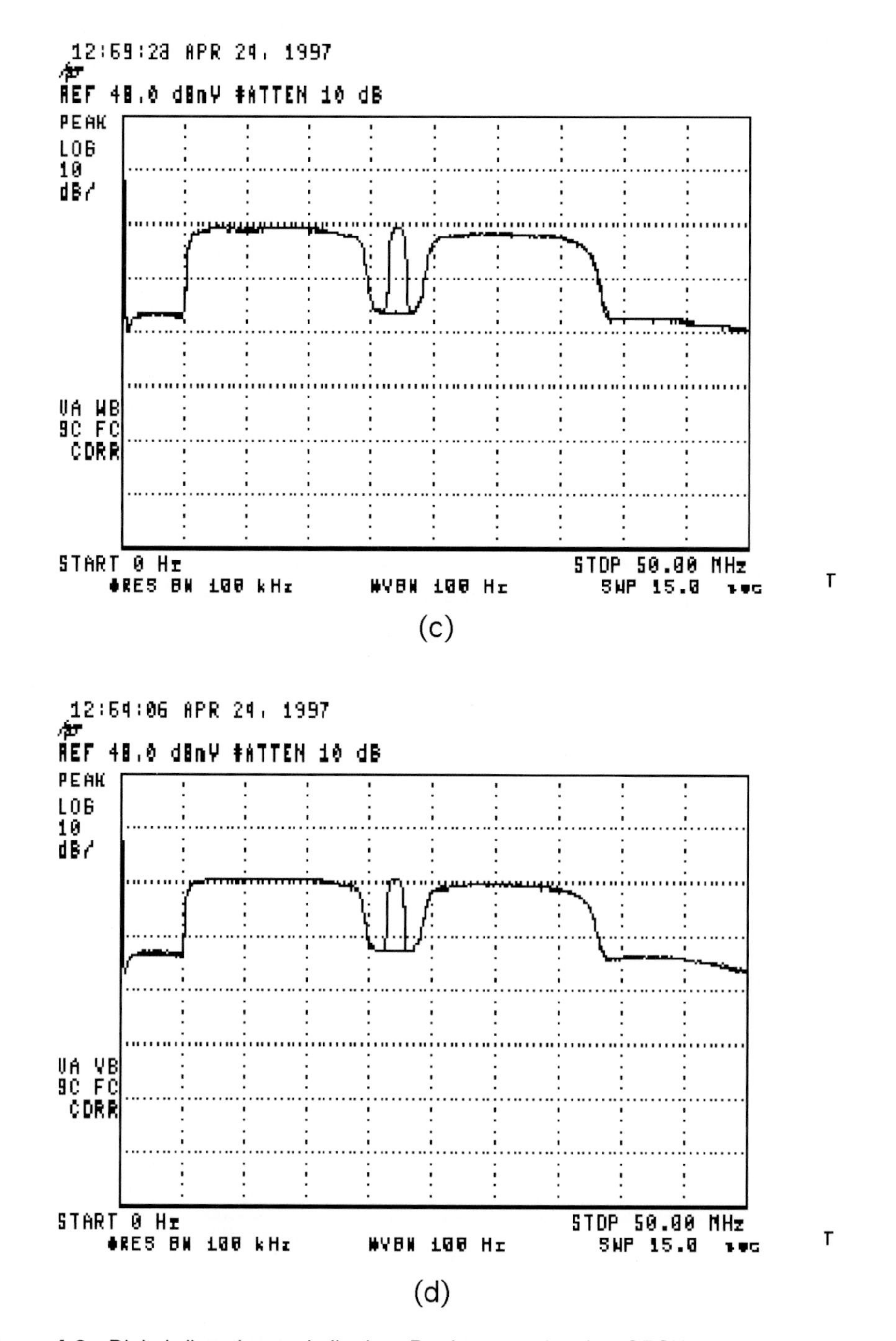

Figure 4-8 Digital distortion and clipping. Dual traces showing QPSK signal turned on and off (c) nominal + 16 dB, (d) nominal + 18 dB

Since clipping creates products at the same frequencies as intermodulations, it isn't easy to determine how much of this "trash" is due to conventional distortion and how much to clipping. We generally concentrate on clipping because such a large quantity of products is created when a laser clips, but in most cases we don't need to make the distinction, as we will see shortly.

Table 4-4 gives the *uncorrected* BER (the bit error rate with no forward error correction) for the QPSK signal in the noise slot during the experiment just described. Note that the BER behaves approximately as predicted by the waterfall curves in Chapter Three (Figure 3.12). From this one could conclude that intermods and clipping cause the same effect as noise.

Table 4-4 BER degradation for QPSK

Test	Input level	C/N	BER measured
4.8(a)	Nom	42.2	$< 1 \times 10^{-9}$
4.8(b)	+10	28.2	$< 1 \times 10^{-9}$
4.8(c)	+16	14.5	4.3×10^{-7}
	+17	12.8	1.9×10^{-6}
4.8(d)	+18	11.7	1.6×10^{-5}

Two sets of tests were done to compare how the BER is affected (a) by white noise and (b) by intermods and clipping. In the first test, white noise is raised below a relatively low-level QPSK-modulated carrier. In the second—a noise notch test similar to the one just described—signal levels are raised until intermods cause errors. Table 4-5 lists the C/N and the carrier-to-intermod value at which the BER became 10^{-6}. For QPSK, the two ratios are the same, within measurement accuracy.[13]

If the same tests are repeated with a 16-QAM signal in place of the QPSK test signal, we see approximately similar behavior: the 16-QAM signal may be 2–3 dB more sensitive to intermods than it is to pure noise. On the other hand,

13. A different modem was used in this test (so that we could also do M-QAM testing). Unfortunately, that makes it impossible to compare the QPSK data between Tables 4-4 and 4-5.

Table 4-5 shows that the BER for 64-QAM degrades much earlier as the levels are raised, than by it does by raising the noise. The difference is due to the closer spacing of the 64-QAM states. Clipping events are of short duration—usually less than a symbol period—but they can make significant changes in signal amplitude. With closely spaced states, the clipping event is likely to cause a symbol to fall into an adjacent state. Figure 4.9 shows schematically the difference between the effect of noise and clipping on a 16-QAM constellation at the same BER. Note that the noise enlarges the constellation points generally, while clipping retains the tight cluster except for some outliers. A carrier-to-noise or carrier-to-intermod measurement essentially records the size of the cluster and will not be very sensitive to the small number of outlying points. This is why it may not correspond well to the measured BER when clipping is the primary contributor.

Table 4-5 Point where BER = $1x10^{-6}$ for QPSK, 16-QAM and 64-QAM

Modulation	Carrier-to-white noise	Carrier-to-intermods
QPSK	23	22
16-QAM	24	27
64-QAM	28	43

The laser transmitter will generate huge quantities of intermod trash if there is appreciable clipping of the RF signal. Consequently, if the RF power into the laser is not well controlled, clipping is likely and the effective noise floor will rise dramatically. Thus the *power control algorithms* used by the return path applications in the home need to be established with care so that no single application can cause its transmitters to "blow away" all of the others. For example, this means that applications cannot start up at high power and then be controlled down.

As we will see in Chapter Ten, return path clipping is *not* likely to be caused by ingress signals; rather, signals intentionally applied will overload the laser. This further emphasizes the importance of coordinated power control for all of the return path applications.

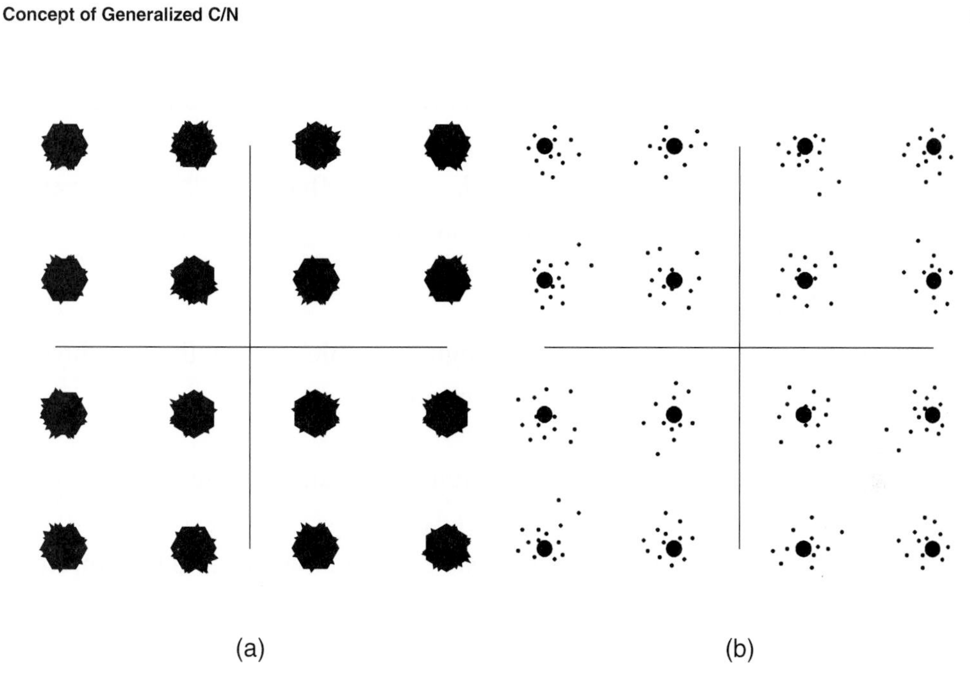

Figure 4.9 Schematic representation of 16-QAM constellations at a BER of 10^{-4} (a) with noise loading and (b) with clipping

Concept of Generalized C/N

We have said that the intermods between digital signals look like noise. Furthermore, we have shown that—especially for lower-order modulation schemes—the BER performance of digital transmissions can be predicted directly from a measurement of the ratio of carrier power to the totality of all the noise-like trash that has accumulated in the noise bandwidth of the signal—both true noise and distortion products. This leads us to broaden our use of C/N to something that includes the intermods. Since both the data signals and the trash are noise-like, we can borrow a term from the early days of radio telephony to describe this generalized C/N—the *noise power ratio* (NPR).[14] The caveat is that for amplitude modulation schemes, the NPR is a good predictor only when the transmission channel (generally meaning the return laser) is in a "well-behaved" region of operation, that is, where clipping does not dominate. Our tests indicate

14. Others have used C/(N + IMN), carrier to noise and intermodulation noise, which appears more proper when written, but is a mouthful to say.

that this becomes a concern only for 64-QAM and higher-order amplitude modulations.

Throughout the remainder of the book we will often use C/N in this generalized sense, which includes well-behaved intermods.

Common Path Distortion

As the bi-directional RF signals pass through the cable system, they go through a number of mechanical contacts. Examples are:

- Push-on G-connectors used between removable amplifier modules and their housings, as sketched in Figure 4.10
- Screw-down seizures to clamp the coax center conductor into the amplifier (also in Figure 4.10)
- Terminators with spring-finger ground contacts.

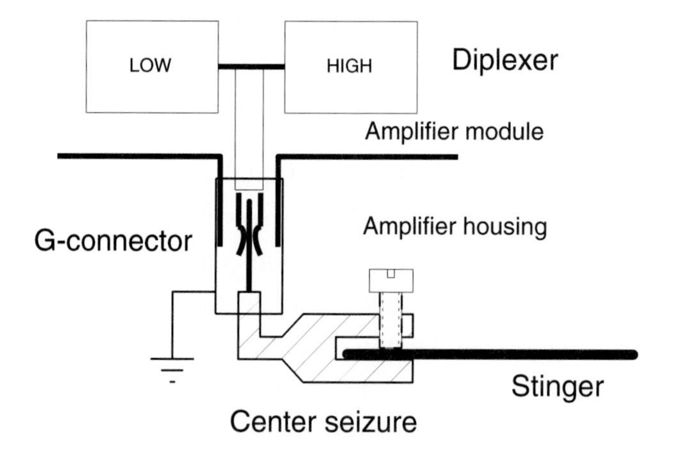

Figure 4.10 Common path trouble spots

If an oxide layer were to build up at any of these points, the contact would develop an electronic potential that functions like a tiny diode. Unfortunately that diode—being a non-linear element in the circuit path—causes the forward signals to mix. The difference beats produced from this mixing fall in the return band. Since the forward and return signals are passing over the same conductor at that point, those unwanted beat signals can combine into the return path spectrum (Figure 4.11). Because this impairment arises at points where both the forward and reverse signals are present, it is referred to as *common-path* distortion. The beats can be significant because the forward levels are high and the return levels low at some of these points. Unlike the other sources of detrimental signals, common-path distortion can be highly variable with no clear correlation to time of day or environmental temperature. That is because the nature of the diode-like contact will vary widely and inconsistently due to several unrelated factors, such as pressure, humidity, and temperature. Note that this is not the same phenomenon as common-mode interference, described earlier.

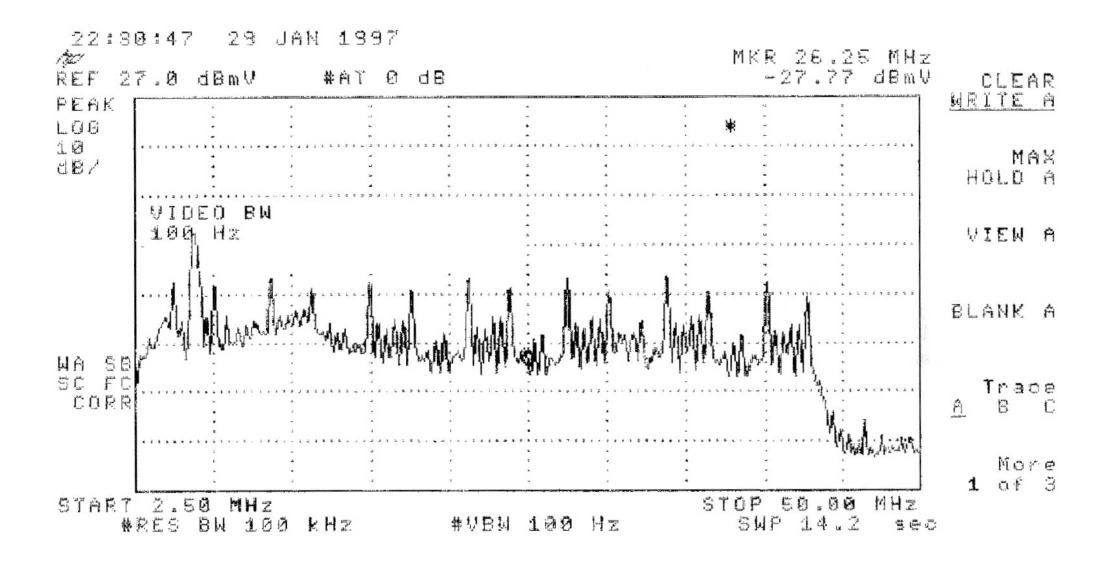

Figure 4.11 Common path distortion in return band

Source: Courtesy of TCI Engineering, Denver, CO.

Since common-path problems occur when these common point contacts become corroded or loose, the preventative is straightforward in principle: making sure that all the contacts are made of similar metals and that they are clean, tight, and water-sealed. Yet, this is difficult in practice because there are so many of these potential trouble spots in a system.

Hum Modulation and Power Supply Spurs

There are two mechanisms by which powering-related interference can show up in the return band. The first is modulation of the return signals by the AC utility power. In the return band this generally means that there is a small amount of unfiltered AC on the DC lines that bias the amplifier ICs. Occasionally this can also result from a defective RF coupling capacitor at the input to an amplifier. The second mechanism is spurious emissions coming from the DC power supply. Any of these problems needs to be corrected at its source (that is, at the offending amplifier or power pack).

Summing-Up...

- While the architecture of the return system tends to accumulate noise, segmentation by HFC designs is a help.
- Fiber links dominate system noise.
- Ingress reduction requires both maintenance attention and technical solutions.
- Modulation schemes of higher order than 16-QAM are sensitive to laser clipping, which may make them difficult to use in the return path.
- Maintenance attention is required to combat noise generated within the system, such as common-mode, common-path, and hum.

CHAPTER 5

Plant Design Considerations

As we have already pointed out, the design of the upstream portion of cable TV systems has been given much less thought to date than has the forward design. This chapter will deal with two consequences of that forward bias: return gain variance and return path loss inequality. The first of these problems has a significant impact on the dynamic range requirements for the return application transmitters in the home. The second has an important bearing on the ingress of noise and interference into the return system. We will describe both of these problems, along with details of what needs to be done to minimize their effects.

Return Path Loss Variance

The total path loss from a subscriber terminal device to the headend varies with time, temperature, and location. In order to understand all of these variations, let's consider an upstream communication signal that is generated in an in-home application transmitter, such as one built into a set-top box or a cable modem. We will trace the path that it takes from the home all the way to that application's receiver in the headend.[1] This is shown schematically in Figure 5.1, where we have emphasized those elements that cause variation in the gain of the path.

1. Be careful to distinguish between the application receiver and the fiberoptic link receiver. In the following discussion, any effects due to the latter receiver will be included in the optical link.

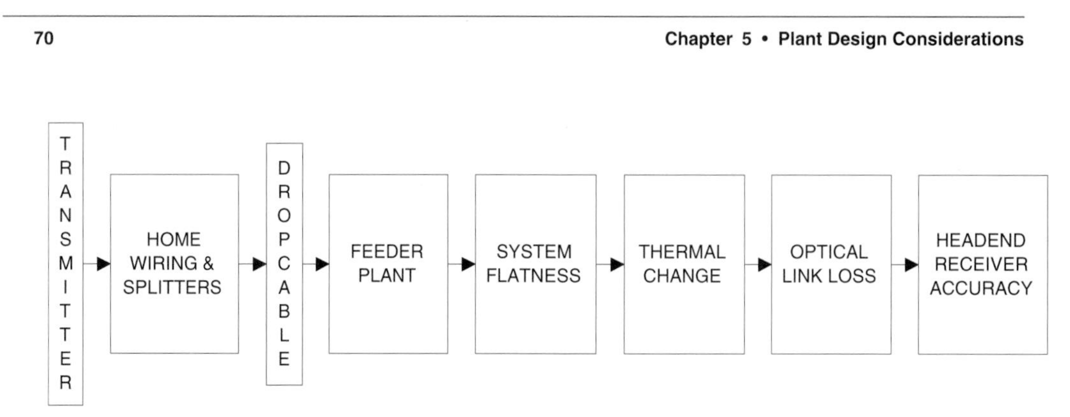

Figure 5.1 Upstream transmission path

Let's discuss each of the elements in Figure 5.1 so we can understand their impacts and decide how to deal with them. We assume that the output level from the transmitter is controlled by a *power control algorithm* that attempts to keep the power received at the headend constant for all of that application's transmitters. For example, the headend controller for the cable modems will command all modems to transmit at a level that arrives at the data communications headend unit at, say 10 dBmV.

Home wiring and splitting loss

When system designers calculate the required signal level at the tap port, they typically take into account the loss through one in-home split, 50 ft of RG-59 indoor cable and 100 ft of RG-6 drop cable. At the return band frequency of 40 MHz, this amounts to approximately 4.3 dB for the in-home loss, if we assume 3.5 dB for the splitter attenuation and 0.8 dB for the cable (Table 5-1). But one splitter accounts for only one TV and one VCR. For the home that is subscribing to several applications, it is easy to envision one or two more TVs plus one modem, as depicted in Figure 5.2. The cable modem in Figure 5.2 would see an additional 50 ft of RG-59 and one more splitter, for a total of 8.6 dB at 40 MHz. Each additional splitter in the path will increase this loss by approximately 3.5 dB.

Three things become evident from this example. First, the cable modem may be transmitting through as little as 3.5 dB or as much as 12 dB of in-home attenuation. The second point, which relates as much to the forward design as to the reverse, is that it is likely that in-home amplifiers (bi-directional) will

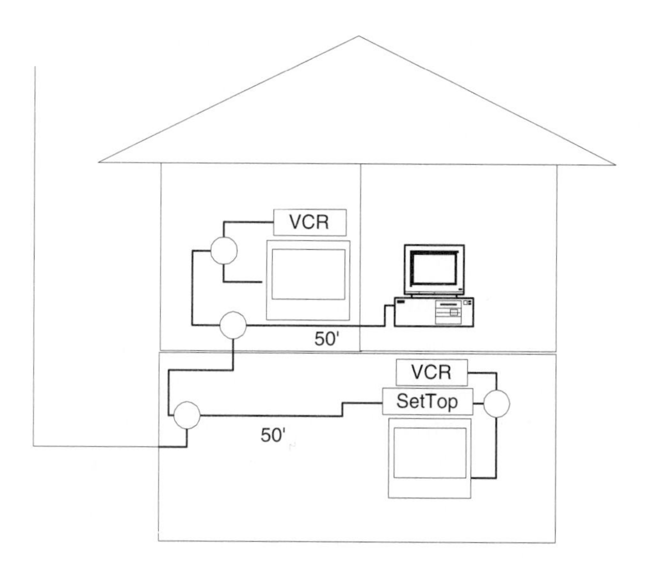

Figure 5.2 In-home wiring

Table 5-1 Attentuation of drop and in-home cable

Frequency (MHz)	Cable Loss (dB/100 ft)	
	RG-59	RG-6
750	7.0	5.6
40	1.7	1.4
5	1.0	0.6

become necessary as the multiplicity of in-home cable applications grows. Finally, one needs to ask the question "What keeps the many terminal devices on the in-home network from 'talking' to one another?" Splitter *isolation* is typically less than 20 dB port-to-port, which means that if the cable modem in Figure 5.2 were to transmit at 55 dBmV, a signal level of 55 – 0.8 – 20 – 3.5 = 30.7 dBmV would reach the back of the second TV. Since the analog TV tuner is expecting a level of –10 to +10 dBmV, it may be overloaded by the modem transmitter. In that event, a high-pass filter will be needed ahead of the TV.

Drop cable

As mentioned earlier, an average drop length of 100 ft is generally assumed by system designers in the US, but actual lengths vary from 50 ft to more than 300 ft. For RG-6 this would imply that the drop loss at 40 MHz could be between 0.7 and 4.2 dB.

Feeder plant

Cable network design produces a variance of return signal level over the transmission path because the tap values are selected to provide proper drop levels at *forward* frequencies. In a tapped feeder line from a line extender (Figure 5.3), the forward signal at each successive tap port in the string is designed to have an approximately equal level at the highest design frequency.[2] This is to ensure a proper forward level to each household. For all other frequencies in the passband, however, the tap output level increases at each successive tap because the loss through the path is lower for every lower frequency. If the difference in level across the forward band gets too large, an *in-line equalizer* is inserted into the main coaxial line to bring the levels within the output specification required by the system design. Typically, however, nothing is done to correct for the path loss differences in the return band, so the upstream path loss will vary widely with every tap. Figure 5.3 shows this by comparing 750 MHz and 40 MHz losses for one tap string.[3] The return path loss variance due to plant design is summarized in Table 5-2.

System flatness

For communications applications that use significant portions of the return bandwidth—such as a cable modem system that dynamically allocates thin

2. Since the signal is attenuated as it goes down the cable, the tap output is kept equal by increasing the amount of signal that is tapped off at each successive tap. Thus the *tap value*, which is the ratio of the through signal to the tapped signal (in dB), decreases as one proceeds down the feeder span.

3. In this example and those following, we ignore the insertion loss of the tap. This makes the diagrams clearer and has little effect on the analysis. Although taps are generally thought to have a "flat" (i.e., non-frequency-dependent) insertion loss, in actuality there is a frequency dependence that is approximately two-thirds that of cable.

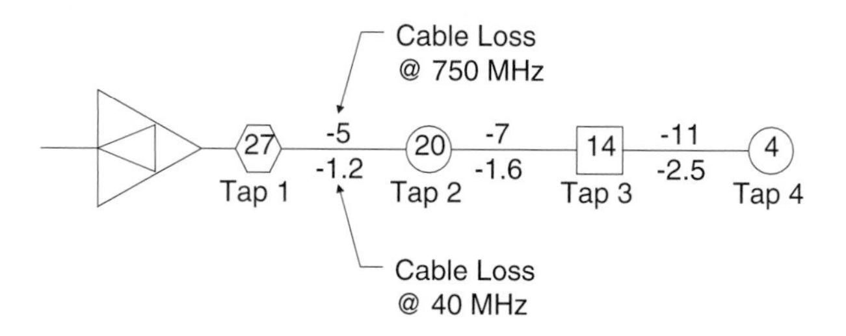

Figure 5.3 Tap selection and cable loss. The number inside the tap symbol is the attenuation through each tap port.

Table 5-2 Feeder plant path-loss variance

	Loss from active to tap port at 750 MHz	Loss from tap port to active at 40 MHz
Tap 1	27 dB	27.0 dB
Tap 2	25 dB	21.2 dB
Tap 3	26 dB	16.8 dB
Tap 4	27 dB	9.3 dB
Difference between tap 4 and tap 1	0 dB	17.7 dB

slices of a 15 MHz band between a number of simultaneous users—it is important to know how flat the return channel response is across that band. A 3–4 dB gain variation across a 15 MHz band is not unusual for short cascades, since there are generally no flatness adjustments in the return portion of HFC amplifier stations. This gain difference must also be compensated for somewhere in the system.

Thermal change

Cable plants are usually specified to operate over the temperature range of –40° to +60°C. There are two thermal effects on the plant gain: cable attenuation increases with temperature, and laser RF efficiency decreases with temperature. With no corrective circuitry, the attenuation change at 40 MHz for four cable spans (3 tapped and one express) is on the order of 6 dB. While closed-loop *automatic gain control* (AGC) is almost unheard of for the return amplifiers, it is possible to correct for most of that change with simple thermal compensation circuits.

The uncooled semiconductor lasers commonly used in the return path tend to have *a slope efficiency*—the change in optical output power for a given change in RF input current—that decreases with temperature, falling fairly dramatically above 60°C. Without thermal compensation, this could amount to another 6 dB change in return gain.

Optical link loss

For a given laser transmitter, the RF output of the fiberoptic receiver at the headend will decrease by 2 dB for every 1 dB increase in optical link loss. Thus the link outputs in a plant that has optical "lengths" varying from 1 dB to 12 dB (3 km to 36 km) could differ by 22 dB. But by adjusting the RF output from the fiberoptic receivers—or by inserting pads on the RF output—it is relatively easy to reduce the variance to 1 dB.

Headend receiver accuracy

As mentioned earlier in this chapter, the application transmitter outputs in each home are controlled from the headend in order that each signal reaches the headend at a prescribed power. This relies on a power measurement by that application's headend receiver. The accuracy of that measurement will naturally affect the actual output power, causing a transmitter to run, say, 2 dB hotter than required because the receiver measures 2 dB low. An accuracy of ±3dB appears to be a cost-effective minimum value.

Total variance

Table 5-3 summarizes these return variances and shows the effect of thermal compensation and optical link adjustment. Nevertheless, if nothing more could be done, we would be left with a need to specify a dynamic range of at least 49 dB for each application transmitter.[4] This applies an undesirable per-home cost to each application.

There is essentially no way to reduce the variance due to in-home wiring and drop length other than installing amplifiers with a compensating amount of gain

Table 5-3 Summary of return path gain variances

System segment	Variance @ 40 MHz (dB) with no corrections	Variance @ 40 MHz (dB) with corrections
In-house cable & splitting	11	11
Drop length	3	3
Feeder plant design	18	18
System flatness	4	4
Thermal change	12	6
Optical link loss	22	1
Headend rcvr accuracy	6	6
Total	**76**	**49**

on the side of every home.[5] However, there *is* a substantial remedy for the remaining large contributor to variance, feeder plant design, as we are about to see.

Return Path Equalization

In order to make the return path loss the same (that is, to equalize the path) between each home and its upstream amplifier, we will need to *add* path loss to most of the homes. This is an unpleasant thought to people in the cable industry, who have made a practice of designing distribution plants to be as efficient as possible by reducing loss. To make the message more tolerable, we need to dis-

4. In the past with only pay-per-view set-top boxes operating on the return path, headend receivers were designed with a large input range, thus making tight control over return transmitter levels unnecessary. As the applications multiply toward a fully loaded upstream system, the available dynamic range of the return band itself must be metered out judiciously. This means that each service needs to have its power under tight control. Since there is a limited carrier-to-noise available, tightly packed systems cannot have some channels operating 10 dB low while others operate 7 dB high. Furthermore, in modern TDMA applications, the levels from all transmitters must arrive at the headend receiver at approximately the same level in order to guarantee proper detection.

5. As we will discuss in Chapter Six, cable telephony systems generally require a side-of-the-home box. In such cases, the incremental cost of an amplifier would be lower. Maintenance and re-balancing of the home amplifier may be a costly headache, however.

cuss an important side-benefit of this loss addition. Then we will discuss the different methods for accomplishing loss equalization.

Ingress reduction

Equalizing the path loss from all homes will reduce ingress into the cable plant. To understand why this is so, let us make the simplifying approximation that RF return band interference levels are equal in all homes (or, alternatively, that there is an equal probability of ingress from any home connected to the system). Figure 5.4 depicts this by showing the system being bathed in a cloud or "ether" of ingressing signals. If noise arrives equally at all tap ports, then the lower-value taps—the ones at the end of the coaxial string—will allow the greatest amount of ingress into the plant. Returning to Figure 5.3 and Table 5-2, we see that if an interference of 20 dBmV strength were omnipresent, then the interference reaching the amplifier port would be –7dBmV from the first tap in the string, but would be +10.7 dBmV from the final tap.

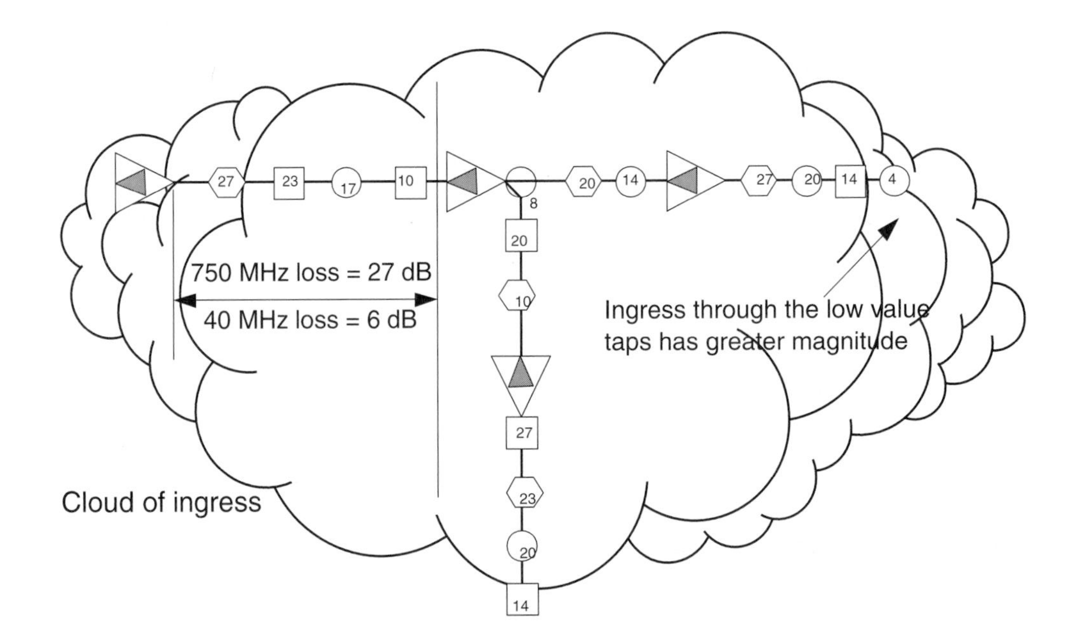

Figure 5.4 Ingress into the distribution plant

An actual HFC node design was analyzed to see how much the ingress could be reduced if extra loss were added to the homes fed by low-value taps. Again, we assumed a uniform background of noise and interference. In the sample area, there was a loss range of 21 dB at 40 MHz. If devices were placed at every tap to equalize the loss—so that the variance would be reduced to within 1 dB—the ingress decreased by 9 dB.

To obtain the full 9 dB improvement in this example would require 20 different equalizer device values, which would be logistically impractical. In order to determine a good compromise between ingress reduction and practicality, we repeated the calculation for various step-sizes of equalizer value. Table 5-4 summarizes the study by showing how many of the 399 taps needed equalizer (EQ) devices and by how much the ingress was reduced in each case. One sees immediately that large step-sizes essentially cause many of the tap values to be neglected, which means they are relatively ineffective in reducing ingress. On the other hand, a 3 dB step-size accomplishes nearly the whole job with only seven different EQ device values. Coincidentally, taps generally also come in 3 dB value steps.

At this point, we can revisit our feeder example and include the effect of equalizers with 3 dB steps (Table 5-5 and Figure 5.5). We can also update the return path loss variance summary to include the improvement in the feeder plant (Table 5-6). The total dynamic range requirement is now reduced to a manageable 34 dB.

Tap-drop equalizers

Our aim is to achieve a uniform return signal attenuation from each tap port to the input of the next amplifier station. There are a number of alternative ways to actually accomplish this. Let's examine the device options.

Where is the equalization device located?

The most immediate answer—installing equalizers in-line with the distribution cable—appears attractive at first, but has several drawbacks. Assume for the moment that the present in-line forward equalizers (used to compensate the forward band for too much cable tilt in the feeder) were to be replaced with circuits that equalized for cable over both bands. Since in-line equalizers appear

Table 5-4 Ingress reduction vs. equalizer step-size

Loss from tap port to active @ 40 MHz (incl EQ)	Number of taps with each path loss				
	Original (no EQs)	10 dB steps	6 dB steps	3 dB steps	1 dB steps
7	1				
8	0				
9	0				
10	0				
11	5				
12	8				
13	2				
14	2				
15	5				
16	10				
17	14				
18	14				
19	15				
20	25	25			
21	25	30			
22	37	45			
23	83	85			
24	56	58	78		
25	56	61	74		
26	4	14	31		
27	12	27	42	120	
28	16	30	63	137	
29	9	24	111	142	399
Ingress reduction (dB)	**0.0**	**3.4**	**6.2**	**7.9**	**9.0**
% of taps with EQs	**0**	**19**	**62**	**91**	**98**

> **Field Trial**
>
> *With the help of Cox Communications of San Diego, we were able to do a test of the capability for ingress reduction with loss equalization. Return step attenuators (page 82) were placed at the tap ports on three nodes. After a complete sweep balancing of the return plant, noise floors were measured before and after installation of the attenuators.*
>
> *The results indicated that equalizers with values in 3 dB increments reduced the ingress level by 10 dB.*

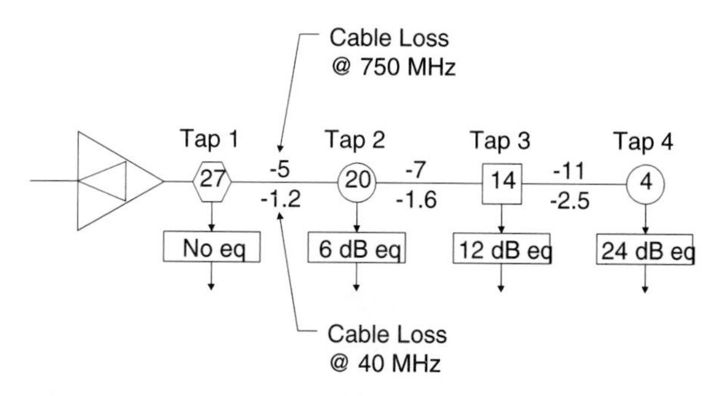

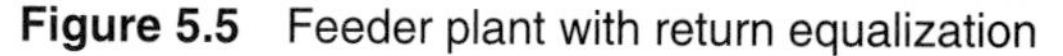
Figure 5.5 Feeder plant with return equalization

Table 5-5 Feeder path loss variance with equalizers

	Loss from active to tap port at 750 MHz	**Loss from tap port to active at 40 MHz***
Tap 1	27 dB	27.0 dB
Tap 2	26 dB	26.8 dB
Tap 3	27 dB	27.0 dB
Tap 4	28 dB	28.8 dB
Maximum difference	2 dB	2.0 dB

* All equalizers are assumed to have a loss of 1 dB at 750 MHz (which is called the insertion loss of the EQ). The equalizer value follows the convention that the value corresponds to the amount of cable for which the equalizer compensates. For instance, a 24 dB EQ compensates for 24 dB of cable (at 750 MHz) and has a loss of 19.5 dB at 40 MHz, including the 1 dB insertion loss.

Table 5-6 Return path loss variances with feeder equalization

System segment	Variance @ 40 MHz (dB) with no corrections	Variance @ 40 MHz (dB) with corrections
In-house cable & splitting	11	11
Drop length	3	3
Feeder plant design	18	3
System flatness	4	4
Thermal change	12	6
Optical link loss	22	1
Headend rcvr accuracy	6	6
Total	**76**	**34**

only rarely at present, the design rules would have to be changed to keep the forward signal considerably up-tilted throughout the distribution (so that more EQ devices would be used, thereby affecting, say 90% of the return paths). Because we have put these devices in the hard-line cable, the path loss between amplifiers in the return band would be the same as for the forward (since equalizers add loss to the lower frequencies). This means that the return amplifiers would need to have the same gain capability as the forward. But that would require nearly a complete re-equipping of the distribution electronics because forward amplifiers are two- or three-stage, but return amplifiers are one-stage only. The amplifier platforms are incapable of that much reverse gain.

It is possible to deploy in-line devices that incorporate diplex filters so that only the return path would be equalized. In order to have the desired effect, there will need to be almost as many of these devices as there are taps, so it appears preferable to incorporate the diplexers and equalizers into the tap itself, if possible. That would avoid the cascaded forward path losses and group delays of the diplexers and the cascaded return path loss of the equalizers.

If we could know for all cases the amount of loss needed for each tap value, it would be a straightforward matter to add the necessary circuitry into a tap. That is not the case, however, as can be seen from Figure 5.6. The use of a directional coupler in the distribution causes the two 10 dB taps to require different amounts of equalization: the upper one needs to compensate for 15 dB of cable, and the lower one sees only 5 dB. This means that if the EQ device is to be incorporated inside the tap unit, it needs to be a plug-in part.

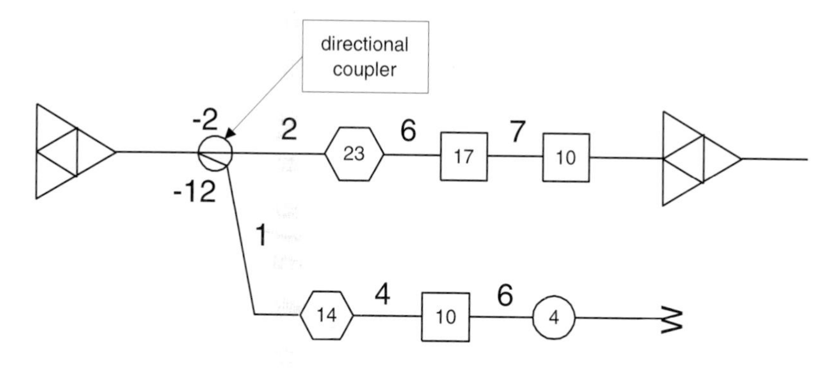

Figure 5.6 Return equalization with in-line directional coupler. The numbers above the cable are losses in dB at 750 MHz.

Incorporating the equalization within the tap is difficult because it is already crammed full with other essential parts, including bulky power-passing components. On the other hand, it is much more economical than the last alternative: mounting two to eight individual equalizers externally on the tap ports (in cylindrical housings with F-fittings, similar to cable traps). Furthermore, external mounting creates a new set of connections that can loosen and corrode over time. For homes that will need a side-of-the-house *Network Interface Unit* (NIU) for telephony or other services, the equalizer could be incorporated at nearly no additional cost. Ingress through the drop cable would not be impeded by an NIU-mounted equalizer, however.

What is the spectral "shape" of the circuitry?

By referring to these devices as "equalizers," we may have led the reader to assume that they had the profile of cable equalizers, with attenuation varying as square root of frequency from the lowest return frequency (5 MHz) to the highest forward frequency, say 750 MHz, as shown in Figure 5.7a. Since the absolute difference in cable loss across the return band is not great, the return loss equalization can be accomplished just as effectively by a device that introduces a fixed, flat loss solely in the return band, leaving the forward essentially unaffected, as in Figure 5.7b.

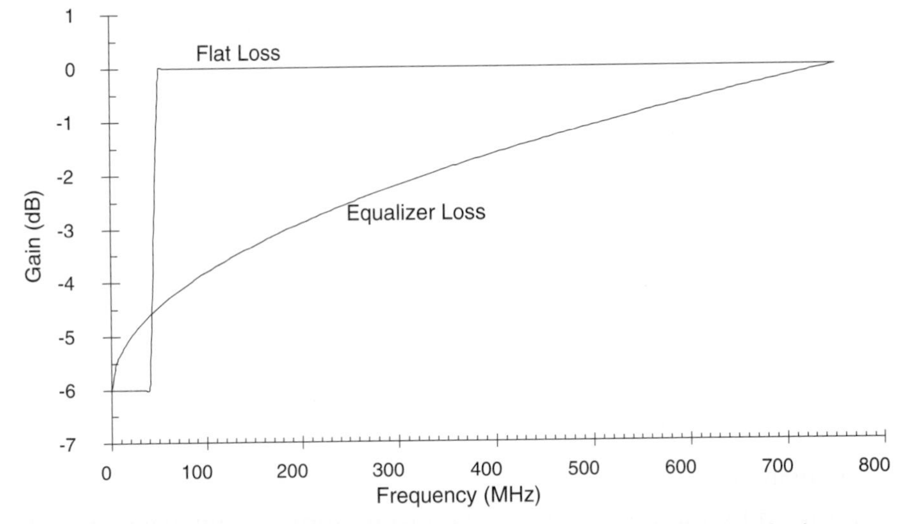

Figure 5.7 Two equalizer types: (a) cable equalizer and (b) return step attentuator

The two approaches are compared in Table 5-7. The device that adds flat loss in the return path, referred to as a *return step attenuator* (RSA), is somewhat more difficult to manufacture, but allows the return path loss to be chosen completely independent of the forward path. For systems with very long drops, the cable-type equalizers will help maintain forward signal levels.[6] In fact, in the ideal case, full equalization at the taps would re-establish the same up-tilt in the forward signal as was present when it exited the amplifier output port. On the other hand, for short drops one would need to ensure that the signal at the drop is not tilted excessively, a problem that occurs only immediately after an amplifier in present designs.

What are the design rules?

For flat loss equalizers:

Find the highest tap value used in the system (call this T_{max}). To find L, the value of loss equalization needed at a tap T, add the loss (at 40 MHz) of the

6. We recently heard of a system with RG-11 drop lengths as long as 200 ft, which corresponds to an attenuation of 12 dB (including in-home cable) and a tilt of 9 dB across the forward band.

Table 5-7 Comparison of equalizer types

	Cable loss	Flat loss (RSA)
Circuit complexity	Low	Moderate
Effect on forward path	Maintains up-tilt	≈ 0.5 dB attenuation
In-line equalizers	Eliminated	Same as before
Fwd bandwidth expansion	Requires replacing EQs	No problem
Rtn bandwidth expansion	No problem	Requires replacing EQs
System design rules	Additional constraints	Straightforward

cable and passives between the upstream active and the tap to the tap value T. Then L will be equal to T_{max} minus that number.

For cable loss equalizers:

A complete model of the plant must be run because the forward and return levels are linked by these equalizers. The in-line forward equalizers need to be removed from the system and the forward plant rebalanced.

Alternatives to Return Loss Equalizers

We have given particular attention to the return equalization concept because it deals directly with two important return path problems: dynamic range requirements and ingress attenuation. Two other approaches to reducing ingress should be discussed as well.

Return blocking filters

It has become common for cable operators beginning the roll-out of two-way services to install high-pass or *window*[7] filters on nearly all of the drops in the service area. These tubular filters block out return band signals coming from those homes not subscribing to two-way services.

7. The window filters block out all of the return band except a window 2–3 MHz wide for analog set-top returns. These allow the addressable set-top communications to pass, while attenuating any other ingress.

Cautionary note:

By selectively adding attenuation to the return system in order to improve the C/N of the entire system, we are causing the in-home transmitter for each application to operate near the top of its power range. This raises two potential problems relating to the isolation and distortion performance of the passive components. Figure 5.8 shows a typical installation, a cable modem communicating with the cable system through an in-home splitter. A TV set, receiving signals from the drop through splitter port 2, is isolated from the modem transmitter by the splitter isolation, which is typically less than 20 dB. If the modem transmits at 31 MHz and has a spurious emission spec of –30 dBmV, then a second harmonic of the transmitter could arrive in Channel 3 at the TV set with a level of –46 dBmV.[a] *Since the TV signals arrive at the set in the range of 0 to 10 dBmV, highly visible interference would be generated by a modem transmitting in the return band.*[b] *It is clear that higher isolation and lower spur specs are required.*

A second concern is distortion in the ferromagnetic components used in splitters and taps.[c] *This causes beats from the return band to appear in the forward spectrum. Only the splitter isolation keeps those beats from the TV set (in Figure 5.8a). For homes connected to one another via the splitter that is in a tap (Figure 5.8b shows a 2-way tap) the downstream signals to the second home will be contaminated by the beats from the first (mitigated somewhat by the drop attenuation from home 1).*

These issues are being addressed by the industry.

a. –30 dBmV transmitter spurious output – 15 dB splitter isolation – 1 dB cable loss.

b. CSO and CTB requirements are typically – 53 dBc to the home, for example.

c. H. Mardia, "Two Way Communications over HFC—The Imperative for New Passives," *Cable Telecommunication Engineers Journal (U.K.)* 19 (September 1996), 42–9.
M.W. Goodwin, "CATV Tap and Splitter Linearity Improvement for Broadband Information Networks, "49th Automatic RF Techniques Group (MTT–IEEE) Conference Digest (1997), 34–38.

While such filters have been very effective in trials, their usefulness is inherently limited. This is because they do nothing to control ingress from the homes that *need* two-way access. If high-pass filters were the operator's only tool for ingress mitigation, he or she would have to live in fear that the new interactive service offerings might be successful! Once the majority of subscribers become two-way, most of the homes will need to be unfiltered, eliminating the barrier against ingress.[8] Operators deploying blocking filters are aware of these shortcomings and view them only as expedients for the roll-out period.

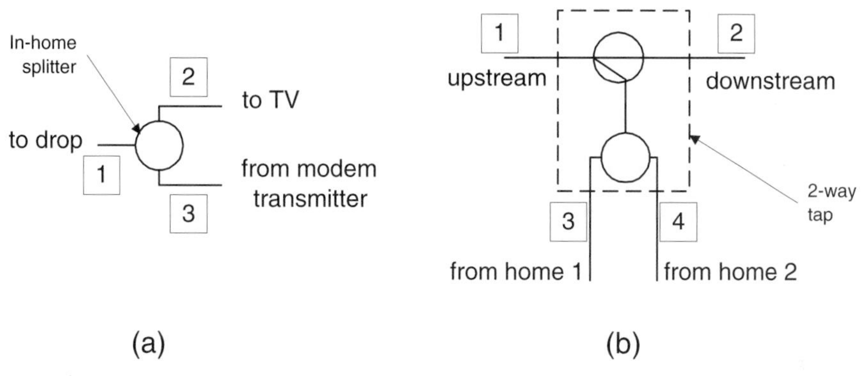

Figure 5.8 (a) in-home isolation and (b) tap isolation

Addressable taps

If fixed filters are not likely to provide long-term protection against ingress, what about filters that are remotely controlled? It is clear that a device that switches in specific values of return path attenuation would be highly effective. It could be used for general ingress reduction and could also aid in tracking down specific interference sources.

Such addressable taps are available commercially, providing control of return path access from the headend. While they presently sell for more than $100 each, the cost should come down with volume deployments and increased competition.

In a sense, however, headend addressability solves only questions of convenience for the operator. A filter still blocks out the return path only when it is active, so many of the performance objections raised about fixed filters are also relevant to addressable ones. With headend control, a subscriber is either blocked full time or open full time (until the access status is changed from the headend). The addressable tap concept would be much more effective if access were authorized from the headend, but controlled from the individual home. That would give maximal ingress blockage with minimal adverse service

8. Furthermore, it has been pointed out that subscribers to two-way services are likely to get "hooked" by their neighbor's good experiences. Thus, even though two-way penetration may be only 20 percent system-wide, it may attain very high rates on individual nodes. In that case, blocking filters will have little effect on the node's ingress level.

impact, at the same time maintaining ultimate override and trouble-shooting capability for the operator.

Demarcation at side-of-home

One way to completely isolate the plant from ingress through the home—either generated in the home or captured by the home cabling—is to avoid connecting that wiring directly to the cable plant.[9] This could be accomplished by providing a *demarcation point* in a small equipment box, possibly mounted on the side of the house. In this concept, the in-home wiring is essentially a self-contained LAN with a bridge at the demarcation point. Circuitry in the side-of-the-home box would receive signals from the home that could be, for instance, baseband digital over twisted-wire pairs.[10] It would translate that information into modulated carrier form before re-transmitting up the cable drop. This provides complete isolation between the home and the cable system. The in-home signaling could be at relatively low levels, since it does not need to be transmitted beyond the demarcation receiver.

Final comment

The previous paragraphs have discussed several ways by which significant ingress reductions can be obtained, but one emphatic statement needs to be made regarding ingress. No design will be effective unless the plant is maintained to high standards. All ground connections and fittings need to be kept tight and corrosion-free. In addition, as we will detail in Chapter Eleven, proper system set-up is critical.

9. J. Terry, Opening remarks, HFC '96 Workshop on High Integrity Hybrid Fiber-Coax Networks, Tucson, AZ, September 1996. (Sponsored by IEEE Communications Society and Society of Cable Telecommunications Engineers.)

10. A number of in-home wiring systems are being developed, such as CEBus and SmartHouse.

Summing-Up...

- Return path gain variance needs to be reduced in order for in-home applications to operate effectively.
- Ingress can be reduced significantly by tap-drop equalization.
- Several alternatives are available for accomplishing this return path loss equalization.
- Issues relating to in-home isolation need to be resolved, either by improving component performance, by filtering, or by using side-of-home demarcation.

CHAPTER 6

Plant Powering

This chapter and the next one on reliability deal with subjects that are not specific to the return path. Even though we are trying to stay focused on return path issues, it is clear that these more general topics of powering and reliability need to be covered because they are critical for providing high, availability two-way plants. To meet these needs, cable operators have had to rethink these matters and to develop some new approaches to them.

How Much Power Is Required

We start by estimating the power required to operate the distribution plant. We'll evaluate at the node level, since that is how powering is segmented. (Fiberoptic cables don't allow powering to be centralized at the head-end without drawing extra cables for power even if that were desirable.)

Assumptions:

- The distribution plant from the node passes 500 homes.
- Population density is 100 homes per mile.
- There are five amplifiers per mile of plant.
- The distribution electronics are half distribution amplifiers (DAs) and half line extenders (LEs).

Typical bidirectional nodes, distribution amplifiers, and line extenders draw 85, 45, and 27 watts each, respectively. For the given assumptions there will be five miles of distribution plant, hence 25 amplifiers. Therefore the power requirement is

$$P_{distribution} = 1 \text{ node x } 85 \text{ W} + 12 \text{ DA x } 45 \text{ W} + 12 \text{ LE x } 27 \text{ W}$$
$$= 85 + 540 + 324 \approx 950 \text{ W}.$$

Note that this comes to 1.9 watts per home passed.

Distribution voltage

This first approximation is an oversimplification, however, because it ignores the power lost in transmission due to resistive losses in the cable. These "I^2R" losses in the cable have three effects. They dissipate energy, they cause the AC voltage to drop through the network, and they heat the cable. All of these effects are compounded when power needs to be supplied to application terminal devices in the home, as will be discussed in Chapter Eight.

The most straightforward method for minimizing these losses is to increase the AC distribution voltage, which reduces the current (the "I" in I^2R). Thus it is becoming increasingly common for the distribution to be done at 90 volts, rather than the traditional 60 volts. We can estimate the effect of distribution voltage by refining our model to include resistive losses in the cable. We will continue to make the simplifying assumption that the AC-to-DC power packs[1] in all of the amplifiers have unity power factor. While this is not always the case, it will keep the calculation from getting too cumbersome, thus allowing us to demonstrate the voltage effect without an overload of math. Essentially, this makes our power distribution network look like a DC circuit. It is unfortunate that the amplifiers don't act like resistors—they represent constant power loads—so a bit of ingenuity is required.

We need to apply a little more detail to our distribution plant by actually locating the 12 DAs and 12 LEs. Let's say that the node has four equal branches, each configured as in Figure 6.1, and let's assume that all of the AC power sup-

1. We need to distinguish between the units that provide regulated (squarewave) AC power to the plant from those units within amplifier stations that convert the AC to DC to run the station. We call the former "power supplies" and the latter "power packs." Unfortunately, this linguistic distinction is not uniformly observed.

plies are located at the node. In this model, legs a and b are express feeders (that is, untapped) of 2,000 ft length[2] each, and legs c, d, and e are conventional tapped feeders of 700 ft length each. The *loop resistance* of the cable, that is, the resistance through the center conductor and return through the outer sheath, is taken to be 1.1 Ω per 1000 ft at the maximum operating temperature.[3]

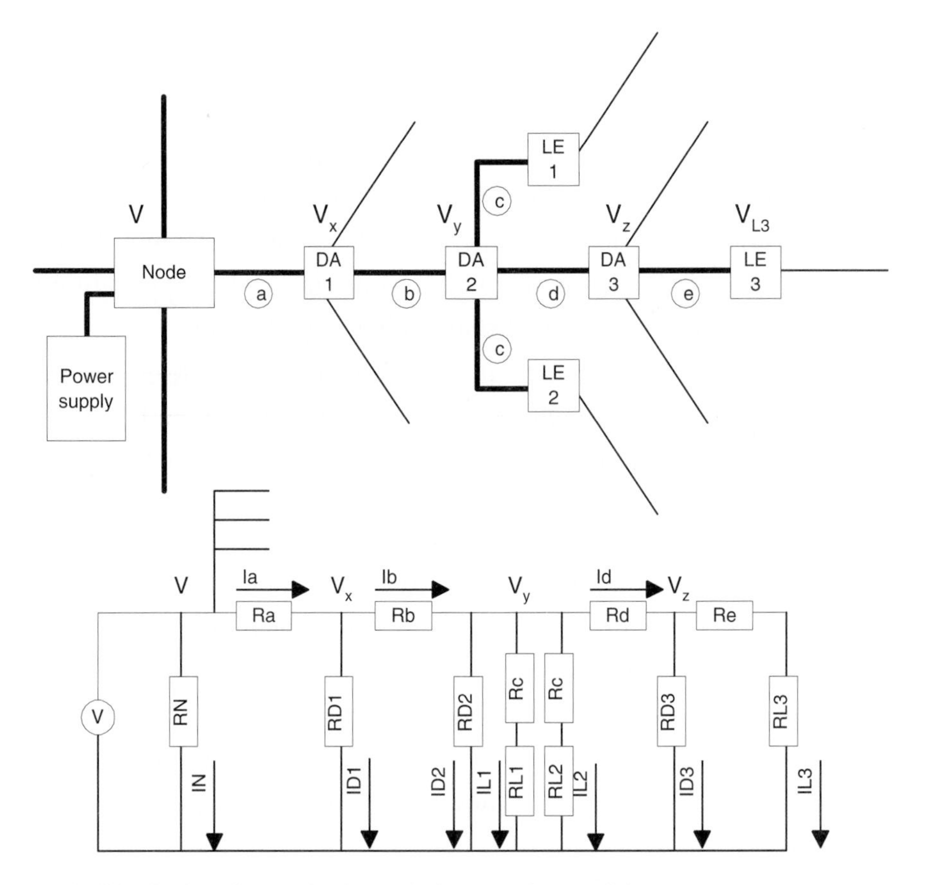

Figure 6.1 Distribution plant and schematic for powering model

2. At 1.5 dB per 100', a typical value at 750 MHz, this amounts to 30 dB for the express portion.

3. Typically, loop resistance values are stated at room temperature and will increase by about 0.2% per °F (or 0.1% per °C).

The analysis is most easily done from the farthest amplifier (LE_3) back toward the power supply. If we call the voltage across that line extender V_{L3}, then the current through that path, I_{L3}, is given by

$$I_{L3} = P_{LE}/V_{L3},$$

where P_{LE} is the power drawn by a line extender (27 W in our model). We then calculate the voltage at the next upstream circuit node point, V_z, by the voltage drop through the tapped distribution cable loss R_e:

$$V_z = V_{L3} + I_{L3}R_e.$$

In a similar way we can work out all of the currents and voltages, going back to the source (Table 6-1):

Table 6-1 Model development

Currents	Voltages
$I_{D3} = P_{DA}/V_z$	
$I_d = I_{L3} + I_{D3}$	$V_y = V_z + I_dR_d$
$I_cR_c + P_{LE}/I_c = V_y$	
$I_c = [V_y - \sqrt{(V_y^2 - 4R_cP_{LE})}] / (2R_c)$	
$I_{D2} = P_{DA} / V_y$	
$I_b = I_{D2} + 2I_c + I_d$	$V_x = V_y + I_bR_b$
$I_{D1} = P_{DA} / V_x$	
$I_a = I_{D1} + I_b$	$V = V_x + I_aR_a$
$I_N = P_N / V$	
$I_{total} = I_N + 4I_a$	

We had to solve a quadratic equation to get I_c. This method has the effect of calculating the supply voltage as a function of the voltage across the last active in the cascade. While this seems to be backward intuitively, it does work. By choosing different values for V_{L3}, we can arrive at commercial values for the supply voltage V. For the parameters we have chosen, the currents and voltages through the system are given in Table 6-2 for both 60 and 90 volts supplied at the node.

Table 6-2 Powering model results

Supply voltage	60 V	90 V
V_x (volts)	47.55	84.04
V_y	37.16	79.27
V_z	35.59	78.57
I_N (amps)	1.42	0.94
I_a	5.67	2.70
I_b	4.72	2.17
I_d	2.04	0.92
ID_1	0.95	0.54
ID_2	1.21	0.57
ID_3	1.26	0.57
IL_1	0.74	0.34
IL_2	0.74	0.34
IL_3	0.77	0.34
Total supply current	**24.09 A**	**11.76 A**
Cable loss	**497 W**	**109 W**
Total power	**1446 W**	**1058 W**

The model indicates that the distribution can be powered from the node at 90 volts without exceeding either the output current rating of the power supply or the current passing capabilities of the amplifiers.[4] The farthest active, LE_3, must be able to operate down to about 78 VAC, but this is not an uncommon capability.

Powering at 60 volts would require multiple 15-amp supplies. If the power were inserted into each of the four output cables from the node, no current-car-

4. Note that the node must distribute nearly 12 A, which means that its power input port must be suitably rated or that some of the power must be inserted into the distribution cables at its output ports.

rying capabilities would be exceeded. The voltage at LE3 would be only 35 volts, which could be marginal, however.

The cable losses raise the power-per-home-passed from 1.9 W to 2.9 for 60 V powering, and to 2.1 W for 90 V. At $0.10 per kWh, the difference in powering cost amounts to $340 per year, or about $2600 net present value for a 10-year life (assuming that rate of energy cost increase is half the prevailing interest rate).

As we will see in Chapter Eight, when the cable network is required to provide power for home application terminals, there is a strong preference for higher supply voltages. In addition, as we will see from the reliability discussion in the following chapter, powering from a single supply may not be a good idea, even when it is feasible.

Powering Architectures

In addition to the voltage issue, the operator needs to choose between two basic schemes for laying out the power distribution. The model we have just discussed has all of the powering concentrated at one point. Our model shows us that this leads to significant power distribution losses—in excess of 10 percent, even in the more favorable 90 V case—and that it is not even feasible in certain cases. Let's discuss the ways to make centralized powering work, and let's also explore the alternative: distributed powering.

Centralized powering

There are two principal reasons for wanting to concentrate powering at one point: the need for periodic maintenance attention, and the increasing difficulty in finding sites for these units. Both of these concerns are aggravated by the need to provide stand-by batteries to back up the network when local utility power is unavailable. The requisite battery packs add greatly to both the space and maintenance demands.

So how can we keep the powering centralized without losing distribution efficiency and without exceeding maximum current or minimum voltage specifications for the plant equipment? As we have seen, raising the distribution voltage is effective, and a number of plants are now being designed and equipped for 90 VAC operation. In addition, some designs are utilizing *express power cable* to

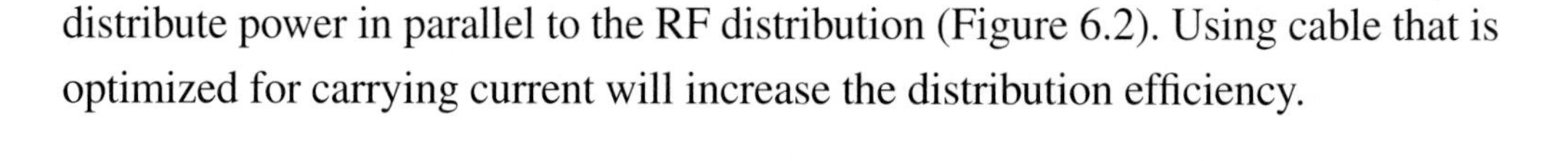

distribute power in parallel to the RF distribution (Figure 6.2). Using cable that is optimized for carrying current will increase the distribution efficiency.

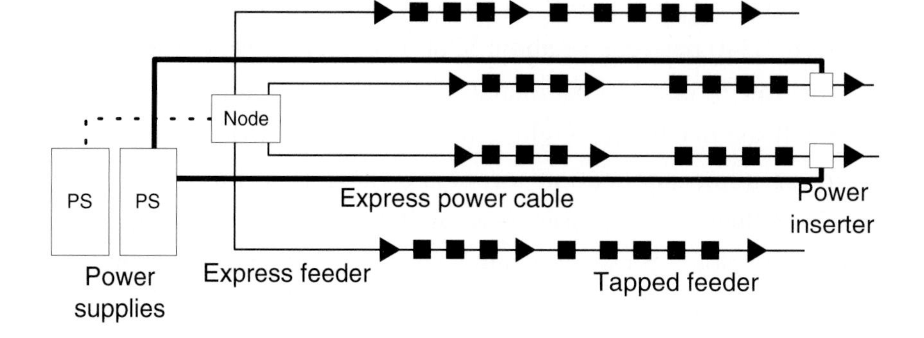

Figure 6.2 Express powering

Returning to our model, we note that if we were able to distribute power with 6,800 ft of parallel cable that had *half* the loop resistance of the coax, there would be a 56 percent decrease in cable power loss for 90 V systems and a 74 percent decrease for 60 V systems. The 10-year net present value of this savings (which would, in part, be applied to the added construction cost) is $420 for a 90 V system and $2,500 for 60 V, using $0.10/kWh. More important, this express powering would make a 60 V system operate with considerably more margin, since LE_3 would be powered at 51 volts.

Distributed powering

As with so many topics in broadband engineering, there are good arguments for the opposite approach as well. By dispersing the powering points throughout the plant, distribution losses can be reduced—without the inconvenience and expense of additional cabling—and availability can be increased. As we will see, one of the most troublesome causes of outages in cable systems is physical damage to the cabling itself, such as automobiles colliding with poles, trees falling on cable during storms and construction mishaps (so-called "backhoe fade").

Providing Backup Capability

When the service requirements include some level of back-up against power interruptions, the tried-and-true 14-amp ferroresonant power supply that has been the work-horse of the US cable industry needs to be augmented with batteries, charger and inverter. The most modest of the uninterruptible units (a 15-amp pole-mount unit) is more than eight times larger than its non-standby equivalent and is twice as heavy—before the batteries are installed. Most of these units are being installed at ground level in pedestal-mounted enclosures or underground in vaults to make maintenance access easier. Because of the size of the units and because of a wide-spread increase in neighborhood hostility toward utility encroachments, it has become very difficult to obtain permits to place power supplies. This problem affects both cost and schedule for construction programs. The use of gas-powered generators in addition, to bridge extended power outages, appears unlikely under such circumstances.

It is unfortunate that there is not much information in the general literature on power outages. The study that is usually referenced was based on data from 1990–92.[5] The data indicate that four hours of battery backup will span 99.96 percent of all power outages since the preponderance of outages consists of short-duration incidents. Nevertheless, those few outages lasting longer than four hours continued for an average of almost four more hours beyond the initial four hours (Table 6-3). Accordingly, if uninterrupted service is required and if independently powered generators are not feasible, then the powering network and the emergency response system need to be thought out very carefully so that these long-duration incidents can be overcome.

Redundant Powering

Distributed powering has the potential for allowing amplifiers to be powered from different cables (from the upstream and downstream directions, for example). In order to realize this potential without destabilizing the AC power system, the amplifier power circuitry must contain some type of *A-B powering*

5. A.L. Black, J.L. Spencer, and D.S. Dorr, "Potential Impact of Commercial Power Quality on Fiber-in-the-Loop Availability," National Fiber Optic Engineers Conference, Book 2, pp. 435–51 (1993).

switch, along with the logic to drive it. The switch would hold station powering on the "A" supply unless that voltage goes below a setpoint, at which time the station would change over to the "B" supply.[6]

Table 6-3 Downtime due to uncovered long-duration power outages.

Backup provided (hrs)	Avg residual downtime after backup (hrs)	Event coverage (%)
4	3.9	99.96
8	3.3	99.97
12	3.1	99.98

Source: from Black, Spencer, and Dorr, "Potential Impact of Commercial Power Quality on Fiber-in-the-Loop Availability." Used with permission.

Summing-Up...

- 90 volt AC powering reduces distribution losses and helps keep current and voltages within rated limits.
- Express power cable can be useful in centralized powering designs.
- Distributed powering is an alternative that needs to be considered.
- The requirements for battery back-up need to be very carefully defined.

6. Note that this is AC power redundancy. DC power redundancy, on the other hand, is based on having two AC-to-DC powerpacks in an amplifier or node station. In DC redundancy, two sources of DC power can be combined directly through diodes without raising concerns about feeding from one source into the other.

CHAPTER 7

Reliability

In this chapter we will present

- a basis for setting a standard for system reliability that is consistent with the service requirements and is competitive with other communications industries,
- a method for predicting the availability of a given network, based on industry data, and
- an application of that method to the principal approaches being implemented to enhance network availability.

Reliability Terminology

We need to define some terms before we start. We are looking for estimates of the *availability* of the distribution network—that is, the fraction of the time that services can flow over the network, often expressed as a percentage. We are interested in systems that are operating in *steady state*, as opposed to the period immediately after construction, when there may be some initial failures (often called *infant mortality failures*) that get screened out rapidly, and as opposed to a late period, when the equipment becomes worn out and needs replacement. In steady-state operation, each type of equipment has a well-defined average failure rate, f (expressed as a fraction or percentage per year). The *mean time between failures* (MTBF) is 1/f years. Each type of failure will take a certain average amount of time to repair (*mean time to repair* or MTTR), during which time the network is out of service. In a year, the *unavailability* U (sometimes called the *fractional downtime*) for each type of equipment will be f x MTTR for that equipment,[1] where the same time unit (for instance, either hours or years) is used for both. The fractional availability A is equal to 1 – U.

What Criteria Are Appropriate?

The individual equipment availabilities can be combined (by a method to be illustrated in the next section) to arrive at the availability for the overall network.[2] Our first challenge is to try to establish what availability is "required" for the network. Efforts to make this determination based on subscriber reactions to outages—such as to become "former subscribers"—have been carried forward by a group of North American cable operators under the auspices of CableLabs, the industry research consortium.[3] Among other findings, this group determined that an outage frequency of greater than twice in a three-month period will cause a significant decrease in customer satisfaction. That work, conducted in 1991–92, focused primarily on the downstream video application. More recently, DAVIC, as part of its standards development effort, has taken the opposite approach, using the service requirements for the specific interactive applications to determine what the network performance needs to be.[4]

A telephone industry guideline for a maximum annual outage of 53 minutes—availability of 99.99%—as put forth by Bellcore,[5] has been applied frequently to

1. Strictly speaking, the unavailability is MTTR/(MTBF + MTTR), which says that the frequency of failure is smaller than 1/MTBF because the system doesn't fail while it is being repaired. For the equipment we are dealing with, the failure rates are so low that this distinction is insignificant. As we will see, the failure rates are at worst a few percent per year, which makes the MTBF = 1/(2% per yr) = (1/.02) yr = (50 yr) x (365 day/yr x 24 hr/day) = 4.4×10^5 hr. Since the MTTR is a few hours at most, the MTBF is 100,000 times greater, thus the quantity (MTBF + MTTR) is essentially the same as MTBF.

2. For a network already in place, the network availability or downtime can, of course, be directly measured. The theory says that the bottoms-up approach, which we describe in this chapter, will give the same result. We will comment on this later in the chapter.

3. S. Bachman et al., "Reducing Outages: A Synopsis of the Current Findings of the CableLabs Outage Reduction Task Force," 1992 NCTA Technical Papers, National Cable Television Association, Washington, DC, pp 243–55 (May 1992).

4. DAVIC Specifications are available on their Web site: http://www.davic.org.

other communications networks. The Bellcore criterion relates specifically to the fiber-in-the-loop architecture, and their 53 minutes do *not* include

- failures of the electric utility power,
- failures of the switching equipment (headend equipment), and
- failures of the cabling from the curb to the home or of the in-home wiring.

The relevance of the Bellcore guidelines to cable plant is debatable.[6] There are many areas where the network envisioned by Bellcore differs from the types of networks we are discussing. Among these are

- Exposure time: Television viewers or Internet surfers tend to be connected for higher fractions of the day than telephone users. This would indicate that the Bellcore service standard might be too low.
- Criticality: Except in the case of primary telephone service (where the operation provides the sole means for communicating in an emergency), the failure of a broadband network is not life-threatening. This would imply that a lower standard could be applied.
- Tolerance level: As new uses of the network come on-line, users may be either more or less forgiving of outages. Thus we should be cautious about predicting what standard would be appropriate in an application universe that we have not yet experienced.[7]

Ultimately, the determination of the network availability requirement is up to the cable operator. That judgment is made in an environment that includes:

5. *Generic Requirements and Objectives for Fiber-in-the-Loop Systems*, TA-NWT-000909, Bellcore, Red Bank, NJ (June 1991).

6. It must be realized that the Bellcore guidelines apply to a telephony architecture that has not been widely deployed as yet. There seems to be a widespread belief that the 53 minute maximum annual outage is achieved by the presently deployed local telephone systems. This is a misunderstanding, and there is no public information to support that view.

7. This caution also applies to evolving telephony systems.

(a) competitive threats for the traditional cable subscriber from alternative providers (telephone companies and satellite or land-based microwave) and (b) an increasing quality consciousness in the subscriber as new interactive services come on-line. Thus, network availability becomes a critical factor for business growth.

Estimating the Availability of Two-Way Plant

We need a model of the two-way HFC distribution plant that will enable us to estimate the overall availability. We will use the model to analyze the potential benefits accruing from various methods for improving reliability, using data from cable operators.

Basics of availability calculations

The availability of the network is calculated by combining the individual availabilities of the network elements (such as nodes, fiber, amplifiers, and taps). First we will be looking at elements that are connected in series only, but later we will look at some parallel (i.e., redundant) configurations, as well. For series elements, the availability of the cascade is the product of the availabilities of the individual elements. For parallel elements, the *unavailability* is the product of the individual unavailabilities. (Recall that if A is the availability and U is the unavailability, then A = 1 – U.)

We need to be able to determine the average availability of the network. Since houses farther from a node will be exposed to longer cascades of the plant equipment, it is natural to expect the availability for individual subscribers to decrease outward from the node. It is therefore useful to think of the service area as N annular rings ("doughnuts") of increasing radius extending out from the node, with constant subscriber density. In this picture (Figure 7.1), it is clear that there will be more homes in the farther rings—the ones that are served by longer cascades of amplifiers. Therefore, the average availability must be calculated by weighting the individual availabilities by the number of homes affected. Mathematically, if n_k is the number of homes in the kth region of the plant and A_k is the corresponding network availability, then the average availability is

$$A = \frac{\sum n_k A_k}{\text{total homes}}.$$

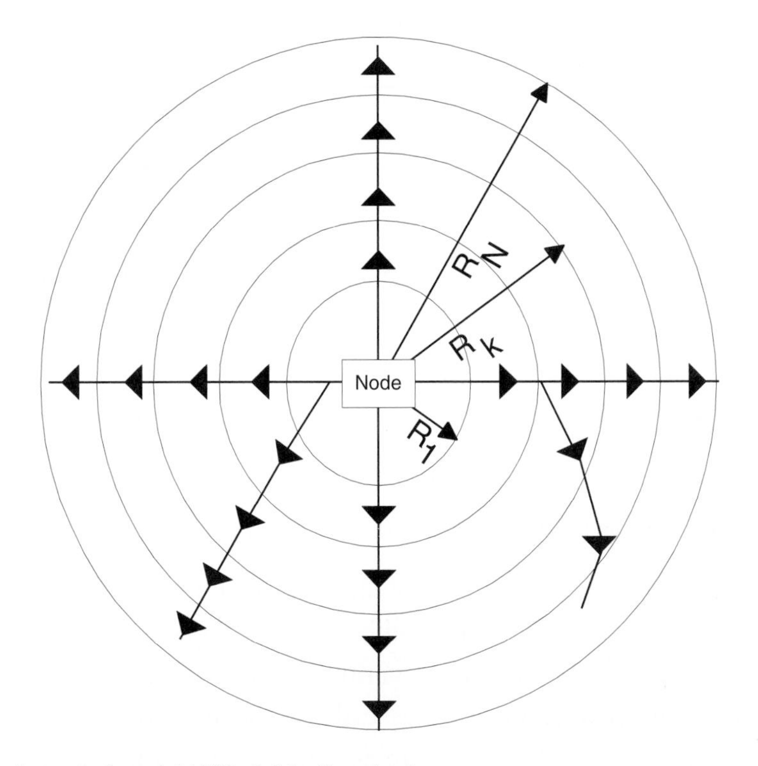

Figure 7.1 Layout of model HFC distribution plant

The network model

We simplify the distribution plant by assuming that

- the subscribers are distributed uniformly around the node,
- the cascade is made up of an express cable followed by N amplifiers with N sections of tapped cable, and
- the amplifiers are evenly spaced.

The express cable runs a distance R_1, as shown schematically in Figure 7.1. The homes in the kth ring are fed by the fiber link, the express span, k amplifiers and tapped spans, and a drop cable, as shown in Figure 7.2. Since we are interested in interactive applications, we need to continue with the return path through much of the same equipment back to the headend. The kth amplifier is a distance R_k away from the node, which in our model is equal to k times R_1.

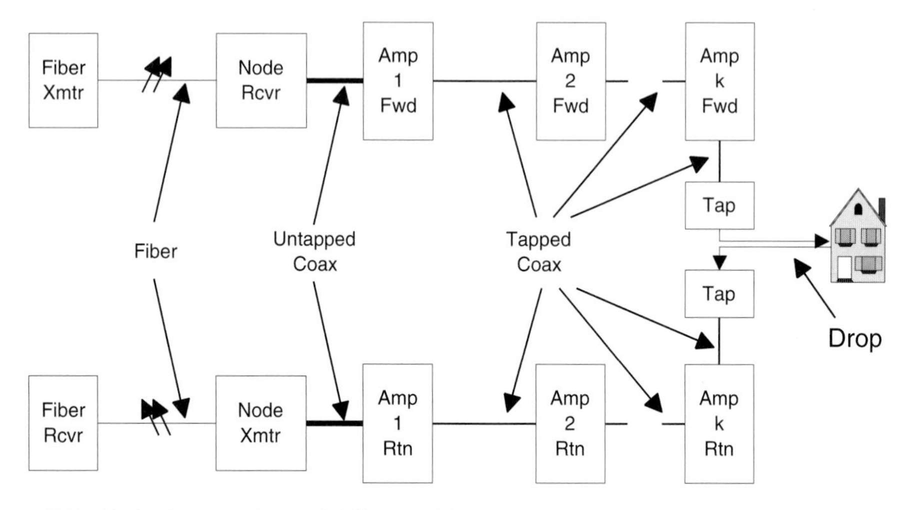

Figure 7.2 Path elements in availability model

The availability will be uniform within any given ring, since the homes in that ring will be served by the same number of amplifiers in cascade. (There will be more than one cascade serving the various homes in a given ring, but all will have the same number of elements in cascade, so the availability will be the same for everyone in that ring.) Since we are assuming a uniform population density, the number of homes in any ring will be proportional to the area of the ring:

$$n_k = (\text{homes per unit area}) \times (\pi R_k^{\,2} - \pi R_{k-1}^{\,2}).$$

But: homes per unit area = (total homes) ÷ (total area) = (total homes) ÷ $\pi R_N^{\,2}$, which (since $R_k = k * R_1$) gives

$$n_k = (\text{total homes}) \times \pi\{k^2 R_1^{\,2} - (k-1)^2 R_1^{\,2}\} \,/\, \{\pi N^2 R_1^{\,2}\}$$

$$= (\text{total homes}) \times (2k-1) / N^2.$$

From Figure 7.2, we see that the round-trip availability of the distribution system for the homes in the kth ring will be <u>the product of</u>:

A_{FX} = Availability of the forward laser transmitter

A_F = Availability of the fiber

A_{FR} = Availability of the forward receiver/amplifier

A_E = Availability of the untapped express coax cable

A_T = Availability of the tapped cable <u>k times</u>

A_{FA} = Availability of the forward amplifier k times

A_{TD} = Availability of the tap and drop two times

A_{RA} = Availability of the return section of the amplifier k times

A_T an additional k times

A_E a second time

A_{RX} = Availability of the return laser transmitter and node return amplifier

A_F a second time

A_{RR} = Availability of the return optical receiver

We also need to factor in the availability of the AC supplies that provide regulated power to the network. Whether the powering is centralized or distributed (see Chapter Six), the number of supplies required for any single leg will be approximately proportional to the length of the leg. We will say that there is always one power supply (PS) at the node plus a number proportional to k:

$$\text{Number of PS} = 1 + (\text{PS per mile}) \times (\text{radius of ring}) \\ = 1 + pk,$$

where p = (PS per mile) x R_1. If we call the availability of each power supply A_{PS}, then the total availability contribution due to power supplies at the kth ring is $(A_{PS})^{(1+pk)}$.

Since all of the rings will have many of these factors in common, we can simplify the formulas by defining the quantity:

$$A_C = A_{FX} \times A_F^2 \times A_{FR} \times A_E^2 \times A_{TD}^2 \times A_{RX} \times A_{RR} \times A_{PS}$$

then

$$A_k = A_C \times A_T^{2k} \times A_{FA}^k \times A_{RA}^k \times A_{PS}^{pk}.$$

Finally, the total average availability will be

$$A = \frac{\sum n_k A_k}{\text{total homes}}$$

$$= \frac{\sum \{(2k-1)(\text{total homes})/N^2\} \times A_k}{\text{total homes}}$$

$$= \sum \left\{ (2k-1) A_C A_T^{2k} A_{FA}^k A_{RA}^k A_{PS}^{pk} / N^2 \right\}$$

$$= (A_C/N^2) \sum (2k-1) A_T^{2k} A_{FA}^k A_{RA}^k A_{PS}^{pk}$$

where the summation is from k = 1 to N.

Reliability data and model calculation

The consulting firm Arthur D. Little has accumulated outage information from ten North American cable operators,[8] which is reproduced in Table 7-1. The MTTR columns distinguish between equipment located in the distribution plant and that which is in the headend, where maintenance access is more immediate.

Table 7-1 HFC cable operation outage data

Component	Failure Rate (%/year)	MTTR in headend (hours)	MTTR in plant (hours)
Fiber transmitter	2.330	1.0	2.5
Fiber receiver	1.396	1.0	2.5
Fiber/mile	0.439		4.5
Trunk Amps	0.514		2.5
LE Amps	0.599		2.5
Splitter/Coupler	0.130		3.0
Tap	0.130		3.0
Hard Connector	0.280		3.7
Coax/mile	0.439		3.5
Power Supply	2.000		2.5

Source: "Failure Modes and Availability Statistics of HFC Networks"

We will use this information to construct availability estimates for our model network. First we need to construct the availability for the three types of cable section (Table 7-2), using the data from Table 7-1.

We will use a fiber length equivalent to 5 dB of optical attenuation, which comes to about 7.8 miles (12.5 km) for 1310 nm lasers. Since the amplifiers in HFC plants are generally either line extenders or distribution amplifiers, we will

8. S. Lipoff, "Failure Modes and Availability Statistics of HFC Networks," HFC '96 Workshop on High Integrity Hybrid Fiber-Coax Networks, Tucson, AZ, September 1996. (Sponsored by SCTE and IEEE Communications Society)

Table 7-2 Cable span availabilities

Element	Express coax		Tapped coax		Tap/Drop	
	Quantity (ft)	Availab'y	Quantity (ft)	Availab'y	Quantity	Availab'y
Cable	2000	0.999999	700	1.000000	125	1.000000
Splitter/Coupler	0	1.000000	1	1.000000	0	1.000000
Taps	0	1.000000	5	0.999998	1	1.000000
Connectors	2	0.999998	4	0.999995	2	0.999998
Availability		0.999997		0.999992		0.999997
Unavailability		3.01E-06		7.60E-06		2.84E-06
Minutes/yr		1.59		4.00		1.49

use the line extender failure rate from Table 7-1. (This is justified because the component count on any single leg of a distribution amplifier is similar to that of a line extender.) For the return amplifiers we have used a failure rate equal to 75 percent of the rate for forward amplifiers. This is based on the approximation that half the failures in the forward amplifier are due to hybrid ICs. The return amplifier has only one IC, while the forward amplifier usually has two. We have used the forward transmitter failure rate for the return laser transmitter, even though the return transmitter is considerably simpler. This is to account for the harsher operating environment for the return transmitter.

We assume two power supplies per mile and R_1 = 700 ft, so p = 2 x (700/5280) = 0.265. For the values we are using, our combined multiplier A_C equals 0.999933.

With these assumptions, we can now calculate the availability for our HFC distribution model for different amplifier cascade depths. This leads to the unavailabilities shown in Table 7-3.

Note that this model *includes* the unavailability due to failures of the tap and drop (3.0 minutes annually). As mentioned previously, the Bellcore fiber-in-the-loop model does not include this part of the distribution in their 53-minute criterion.

Table 7-3 Downtime (minutes/yr) for model HFC plant

N =	1	2	3	4	5	6	7
Fwd fiber link	12.8	12.8	12.8	12.8	12.8	12.8	12.8
Express cable	3.2	3.2	3.2	3.2	3.2	3.2	3.2
Fwd amplifiers	0.9	1.6	2.2	2.8	3.4	4.0	4.6
Rtn amplifiers	0.7	1.2	1.6	2.1	2.6	3.0	3.5
Tapped cable	8.0	14.0	19.6	25.0	30.4	35.8	41.1
Tap & drop	3.0	3.0	3.0	3.0	3.0	3.0	3.0
Rtn fiber link	13.6	13.6	13.6	13.6	13.6	13.6	13.6
AC pwr supplies	3.8	4.4	4.9	5.5	6.0	6.6	7.1
TOTAL	45.9	53.6	60.8	67.9	74.9	81.9	88.8

Some observations can be made from the results in Table 7-3. The first is that the tapped cable is a major factor, with increasing importance as the cascade lengthens. This is consistent with the experience of operators, and it highlights an area that merits attention. The second is that the fiber links are large contributors, as well. When we break the links down to transmitter, fiber, and receiver (Table 7-4), we see that it is the fiber that is the major contributor, not the opto-electronics. In a Bellcore study,[9] the failure rate for optical fiber cable is given as 0.0717 percent per year per mile, which is only 16 percent of the number compiled by Arthur D. Little. Cable operators and telephone companies use the same fiber cable and attach to the same poles. While somewhat more of the telephone fiber is buried, there is little reason to think that the difference is real. As can be seen from Table 7-4, there is nearly an 8 min/yr difference in outage when the two failure rates are compared. If nothing else, this emphasizes the fact that we are merely estimating, with only modest accuracy.

9. *Residential Broadband Quality of Service (QoS)*, SR-3770, Issue 1, Bellcore, Red Bank, NJ (December 1995).

Table 7-4 Analysis of fiber link unavailability

	Arthur D. Little data		Bellcore fiber data	
	Availability	Downtime (min/yr)	Availability	Downtime (min/yr)
Forward transmitter	0.999997	1.4	0.999997	1.4
Forward fiber	0.999982	9.3	0.999997	1.5
Forward receiver	0.999996	2.1	0.999996	2.1
Total	0.999976	12.8	0.999990	5.0
Return transmitter	0.999993	3.5	0.999993	3.5
Return fiber	0.999982	9.3	0.999997	1.5
Return receiver	0.999998	0.8	0.999998	0.8
Total	0.999974	13.6	0.999989	5.8

The Impact of Reliability Enhancements

We can now use our model to examine the effects of various methods for enhancing the reliability of the network.

Fiber link redundancy

First let's see what happens when we use two separate pairs of fiber links, routed along two different paths from the headend to the node. Since these are parallel, independent systems, the combined availability is determined by multiplying the unavailability fractions (probabilities) for each and subtracting from 1.[10] In our model (using the Arthur D. Little fiber outage data), this would mean downtimes of

$$D_{FWD} = [12.8/(365*24*60)]^2 = 5.9 \times 10^{-10} \text{ or } 0.0003 \text{ minutes/yr}$$

and

$$D_{RTN} = [13.6/(365*24*60)]^2 = 6.7 \times 10^{-10} \text{ or } 0.0004 \text{ minutes/yr.}$$

10. As noted at the beginning of the chapter, the unavailability of parallel elements multiply, since the overall system will be available unless both of the links are down simultaneously.

It is clear that both of these lead to availabilities essentially equal to 100 percent. The net effect is that the 26.4 minutes of annual outage attributed to the fiber links in our original model are now eliminated. For the N=5 example, this amounts to a 35 percent outage reduction.

It is important to point out that this analysis has assumed that the links are completely independent. Since the most common failure mechanism for fiber is to be severed by some external event, such as a truck hitting a pole, the redundancy will have most value only if the fiber routings are completely separated. Redundant fibers in the same cable sheath are likely not to have independent failures, which would mean that only the 7.8 min/yr of fiber transmitter and receiver downtimes would be eliminated.

When we discuss network architectures in Chapter Twelve, we will see examples of multiple levels of fiber distribution, such as primary rings feeding subsidiary fiber distribution points. The analysis just done shows that if the ring is bidirectional (or otherwise redundantly routed), then that entire layer of distribution will cause negligible outage time. This is an important result.

Reduced fiber counts

As can be seen from Table 7-1, the highest MTTR is for fiber cable. This is because, as just mentioned, fiber failures generally occur only when the cable is snapped by an external accident.[11] That means, however, that service restoration usually means having to splice all of the fibers in the cable. Thus, restoration time is by-and-large proportional to the fiber count. In our example, if halving the fiber count also halves the MTTR, then the annual outage time would be reduced by 9.3 minutes, or 12.4 percent in the N = 5 case.

11. Despite the fact that the fibers are made of glass, fiber cables are specially constructed for high pull strength, due in part to incorporation of either metallic or non-metallic tension members. Two experiences of US cable operators are of note. The first relates to a hurricane in the Northeast that broke a section out of a pole, leaving the cross-arms and top section suspended by the intact fiber cable. The second resulted from a garbage container truck backing up through a utility span with its container in the raised position, thus pulling down all of the cables. Again the fiber cable was left intact, but when the fire department arrived to deal with the severed electric wires, they chopped through the fiber cable on both sides of the truck to free the truck.

Status monitoring

There are two principal ways in which status monitoring can reduce outage time: (a) by alerting the operator to incipient failures so that actions can be taken before an outage occurs and (b) by pinpointing the nature and the location of a failure when it does occur, thus reducing MTTR. From the information in Tables 7-3 and 7-4, we can see that the electronic equipment accounts for 13.8 minutes of annual outage time (N = 5 case), which is 18 percent of the total. Of this, 2.2 minutes (2.9%) accrues from the distribution equipment in the headend, 5.6 minutes (7.5%) from the node equipment, and 6.0 minutes (8.0%) from all of the amplifiers.

This confirms the widely held belief that the priority for implementing status-monitoring equipment should be the node stations. One transponder in the node has as much impact on availability as does deployment of monitors in all of the amplifiers. We will discuss the role of status monitoring in more detail in Chapter Thirteen, as part of our treatment of network management.

Standby power

The AC power supplies account for 6.0 minutes of the annual average outage in the N = 5 cascade. Our analysis has a weakness, however, in not including electrical utility outages, which could cause all of these supplies to fail—unless they are backed up by standby power. There is no assurance that the subscriber's residential power will fail at the same time as the cable distribution, since the connectivity of the electric utility grid may not be identical to that of the cable operation. Furthermore, as we will discuss in the next chapter, if the cable operator's service offering includes emergency telephone service (Dial 911), cable powering must operate independent of the local electric utility in an emergency.

If standby powering is required, then we need to know how long an electrical outage period must be spanned by the backup system. As noted in Chapter Six, some electric utility power outages can exceed four hours (see Table 6-3). For the most part, these long-duration outages occur either during natural disasters or in remote rural areas. Accordingly, most systems can cover outages with four-hour backup capacity.

Summing-Up...

- There are no well established availability standards for broadband networks.
- Customer expectations for service availability will depend on the services offered and on perceptions of competitive offerings.
- HFC networks are capable of providing high availability.
- Redundant fiber paths essentially eliminate downtime due to fiber outages.
- Reducing fiber counts can have a significant positive impact on availability.
- Reliability models can be useful for choosing between investment alternatives, but do not have sufficient accuracy for predicting absolute outage times.

CHAPTER 8

Application-Oriented Design

This chapter answers a number of "how-to" questions that are central to delivering new applications over the return path:

- What are the methods for estimating how many subscribers can be served by each of the services, for given bandwidth allocations?
- What measures can be put in place today that will permit the network capacity to be increased as the subscriber "take-rate" for services rises?
- How much equipment will be required in the plant and in the headend?
- What are the best ways to segment the return band spectrum?

We will discuss the particular characteristics and requirements for each return path service application.

Interactive Applications

Table 8-1 shows the applications that utilize the return band, along with a representative list of suppliers of the products associated with those applications, and standard-setting bodies, where appropriate.

Table 8-1 Return applications

Application	Suppliers (representative listing)	Standard-setting bodies
Cable modems	Motorola	MCNS
	Bay Networks	DAVIC
	Zenith	IEEE 802.14
	NextLevel	
	Scientific Atlanta	
	US Robotics	
Cable telephone	Motorola	
	ADC Telecom	
	Nortel	
	Philips	
PCS	Lockheed Sanders	
	PCS Solutions	
Status Monitoring	AM Communications	CableLabs
	Superior Electronics	(coordinating)
Digital converters	NextLevel	
	Scientific Atlanta	
	Pioneer	
	Thomson	
Analog converters	NextLevel	
	Scientific Atlanta	
	Pioneer	
	Zenith	

Multiple access and polling

In at least one respect, cable delivery of interactive services to the home is fundamentally unlike conventional wireline telephony. In cable delivery, all of the customers in a serving area[1] share the distribution medium between the signal origin point (headend) and their homes. In telephony, only a portion of the medium is shared (Figure 8.1).[2] In the *digital loop carrier* systems that are commonly deployed for local telephone distribution, a fiberoptic cable connects the *central office* (CO), where the main switch resides, with a *remote terminal* (RT), where individual signals are separated and sent to homes on dedicated twisted-wire pairs. The fiber portion is shared in a *time division multiplex* (TDM), which assigns time slots on a high-speed digital carrier to users as they make or receive calls. Multiple calls are time-interleaved on the fiber. In the RT, the TDM data stream is demultiplexed, and each call is routed to the line interface card that serves the appropriate home.

For services delivered over cable, the entire path is a shared medium. This means that in order for specific communications to take place between individual units, all traffic must be controlled by protocols incorporated in all of the equipment connected to the cable. Note that for the conventional downstream video services over cable, these protocols are not needed since the same signal goes to every home (or at least to every tap).

In many ways, two-way communications over a cable system are like those on a *local area network* (LAN), and several of the cable protocols are adaptations from the LAN world. Communication protocols divide into two broad categories: *polling* and *multiple access*. In polling, a central unit has complete control over the communication links. A familiar example is the addressable controller for set-top converters. The controller sends out a message to a converter with a specific address to establish authorization or to request usage information. The converters transmit messages only when they have been asked to. This "speak only when you are spoken to" protocol ensures that only one conversation can take place at a time. This means that one narrow communication channel can be used for a large number of converters. Such efficient use of bandwidth is possible because most of the information being communicated is not extremely time-sensitive, so the

1. The area fed from one fiberoptic node.

2. For simplicity, the figure omits homes fed directly from the headend or central office by coax or twisted-wire pairs.

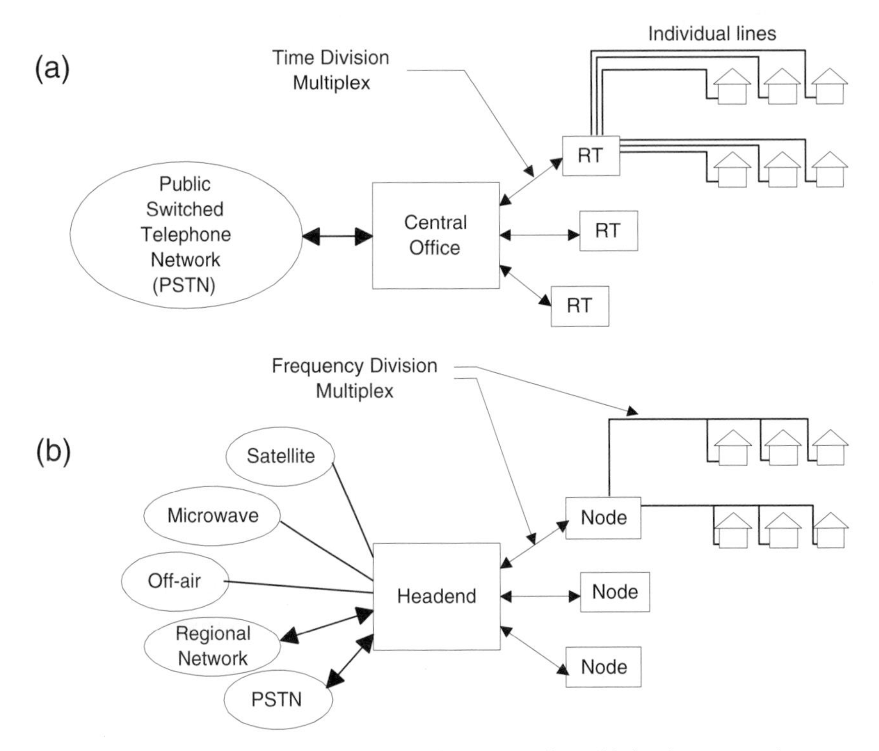

Figure 8.1 Basic distribution architectures. (a) conventional telephony and (b) cable services

controller can "make the rounds" sufficiently quickly to access all of the information held in the converters adequately.[3] Hence the name "store and forward."

Polling is also used to access the information in the transponders that monitor the status of plant equipment. In this case the protocol has to allow certain upstream communications to take place spontaneously, for instance to notify the network management system of an impending or actual equipment failure. One way to make this possible in a polling system is to have the central controller issue frequent requests for "change of status" information, addressed to all transponders. In the event that more than one transponder needs to send such a message in the same interval, a simple method for handling contention needs to be invoked,

3. There are also special techniques used for "fast polls" that do a rapid scan of all stored information, when necessary.

such as for each to back off for a random number of milliseconds before re-trying. These methods have been shown to be effective, even in the not uncommon condition where a failure in one part of the network causes multiple alarms.

In applications where multiple simultaneous communications are the rule and where time delays are detrimental, however, polling is not sufficiently effective. Those situations call for more elaborate protocols that enable multiple users to share the common medium at the same time. Two of these are time division and frequency division (Figure 8.2).

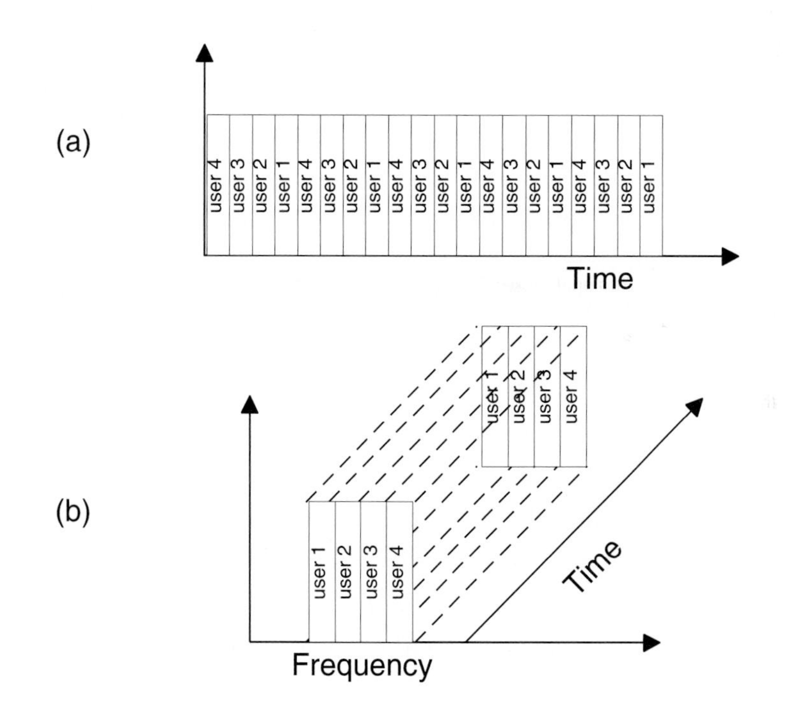

Figure 8.2 Multiplexing schemes. (a) time division and (b) frequency division

In *time division multiple access* (TDMA), N users share the data channel by having their data interleaved in time, so that every N^{th} data packet belongs to a given user (N=4 in Figure 8.2a). In *frequency division multiple access* (FDMA), each user is assigned a carrier frequency within the overall band allocated to the application. Although the details of how time or frequency slots are assigned is beyond the scope of our discussion, it is important to note that this is done dynamically. The subscriber at 123 Elm Street does not "own" time slot 17 in his node's TDMA system. When that subscriber logs on to an interactive service, the controller for that service makes an assignment from the slots that are (a) not in use and (b) adequately noise-free to permit communication. In some systems, the assignment may even change during the session if conditions change.

A variant on FDMA is the access scheme called *orthogonal frequency division multiplexing* (OFDM). In OFDM a relatively wide frequency band, say 6 MHz, is subdivided into numerous individual frequency tones. The word "orthogonal" implies that there is no overlap between any of the tones,[4] so that they can be thought of as being completely independent of one another. The feature of OFDM is that it maximizes the utilization of the wide channel, even in the presence of interference, by subdividing the signal and carrying it on several of the subcarriers. If an interfering signal is present, the signal is moved from the affected subcarrier to another. Thus the clean portions of the 6 MHz are always being utilized. As an example, one cable telephony system assigns 64 kbps voice channels to tones spaced only 18 kHz apart, using 32-QAM modulation.[5] At five bits per symbol and using two tones per voice channel, this implies a rate of less than 7 kilo-symbols per second. These low-rate transmissions should be resistant to impulsive noise because the detector bandwidth can be small, whereas the impulse energy is broadband. The system is claimed to have a capacity of 240 voice channels within the 6 MHz. If ingress causes the channel to degrade, fewer connections can be accommodated. Since it is critical that the basic waveforms remain orthogonal, certain channel impairments, such as phase shifts, can be troublesome. OFDM has been used extensively for voice radio communications in Europe.

Obviously, it is possible to create a hybrid protocol that time multiplexes on a number of frequencies, as well. This is often used in *code division multiple access* (CDMA), which is being deployed for certain cable return applications.

4. Specifically, the overlap integral—the integral of the product of the two wave functions over a complete cycle—is zero. This is similar to the non-interfering I and Q signals in QPSK or M-QAM.

5. T. Schieffer and G. Hutterer, "Maximizing the Return Path in HFC Plant," *CED Magazine*, September 1996.

In CDMA, the individual digital bits are subdivided into a number of sub-elements called *chips* such that the bits generated in each session have an assigned pattern. The receiving equipment looks only for bits that have the desired pattern and ignores all others, thus permitting multiple simultaneous sessions on one communication medium. CDMA is robust against reflections within the system, since any reflected signal that is displaced by more than one chip time period appears as noise. The CDMA concept is readily understood by an analogy[6] to a cocktail party that you are attending, where no one is speaking English, which is the only language that you understand. You are unable to distinguish the conversations from a general background of noise. If, at some moment, one English conversation were to begin, you would have no trouble understanding what was being said, even in the presence of so many other simultaneous conversations. The CDMA detector plays a similar role in separating the message with its particular code from a collection of other messages that appear to be merely adding to the noise floor.

Latency

Cable networks are larger in extent than the typical LAN, which is usually contained within a single building or a small campus of buildings. The fact that it takes an appreciable amount of time to communicate across the cable network has some impact on the access protocols. Signals propagate at 88 percent and 69 percent of the speed of light in coax and fiber, respectively.[7] Thus a signal takes 87 μsec to traverse a path consisting of 1.5 miles of coax and 10 miles (≈ 6 dB) of optical fiber, while signals generated from close-in sources will suffer very little delay. If that signal is being carried in a 10 Mbit/sec stream, then 870 bits (109 bytes) will have been transmitted before the first bit is received. For reference, this is nearly twice as long as one 56-byte ATM packet (see Appendix C). Properly designed systems are able to accommodate these delays, but it is a

6. Qualcomm web site: http://www.qualcomm.com.

7. Note that in this case "light" goes slower than "electrons." This is because the dielectric constant of cable dielectric is closer to unity (vacuum) than is the dielectric constant for glass.

complicating factor. Certain services, such as voice telephony and "twitch" video games, are very sensitive to latency.

Estimating Traffic Capacity

Before beginning a system design, the most important number that an operator needs to determine is the number of homes passed per fiberoptic node (HPN). That number has a major effect on construction planning and cost, since each two-way node with associated fiberoptic headend equipment will cost approximately $8,000–$15,000, plus fiberoptic cable and installation. For that reason, one doesn't want to undersize the area of node coverage. On the other hand, one needs to ensure that the service-carrying capacity of the node will be sufficient for all of the new market opportunities that will arise. Once the HPN criterion has been established, the plant design worked out, and the construction completed, the physical plant will always carry some remnants of that decision—although, as we will see in the next section, some design approaches do allow nodes to be subdivided as future needs require.

In many cases it is the return band capacity that limits the number of subscribers that can be fed from a node. The method for estimating this capacity for each service starts by determining N_{simult}, the number of simultaneous upstream communications that can be handled for that service application at one node. N_{simult} can be estimated in a two-step process. First, the total bits per second (bps) capacity of the node for this application will be the bandwidth allocated for the application (BW) times the bandwidth efficiency (bps/Hz) of the application. Second, that capacity will be fully utilized when the maximum number of simultaneous users is each transmitting (at the application's bit rate). This can be expressed as:

$$\text{Capacity} = [\text{bps/Hz}] \times [\text{BW per node allocated for the service}]$$

$$\text{Maximum utilization} = N_{simult} \times [\text{bps per user}].$$

When these two expressions are equal, the capacity will be fully utilized. This allows us to solve for N_{simult}, as follows:

$$N_{simult} = \frac{(\text{bps/Hz}) \times (\text{BW})}{(\text{bps/user})} .$$

For convenient reference, we have transcribed the theoretical bits/sec per Hz information from Chapter Three into Table 8-2. Remember that practical systems may achieve only 80 to 90 percent of these values.

Table 8-2 Theoretical bits/sec per Hz for various modulation types

Modulation Type	bps/Hz
PSK, FSK, ASK	1
QPSK	2
16-QAM	4
64-QAM	6
M-PSK	Log_2M
M-QAM	Log_2M
M-FSK	$1/(\text{Log}_2\text{M})$

As an example, if 15 MHz is allocated to a cable modem system that is based on QPSK modulation upstream (at 90% of 2 bps/Hz = 1.8 bps/Hz) and that provides 500 kbps service to each user, then

$$N_{simult} = 1.8 \text{ x } (15 \text{ x } 10^6) \div (500 \text{ x } 10^3) = 54 \text{ users/node.}$$

We can calculate the capacity of the node for this service (HPN) from N_{simult}:

$$N_{simult} = (\text{HPN}) \times (\text{Subs/HP}) \times \begin{pmatrix}\text{Take-rate for}\\ \text{this service}\end{pmatrix} \times (\text{Utilization factor})$$

where the second factor is the cable penetration, and the third is the service penetration (as a fraction of cable subscribers). The utilization factor is the maximum fraction of the service subscribers that are actually connected at any one time. By rearranging:

$$\text{HPN} = \frac{N_{simult}}{(\text{Subs/HP}) \times (\text{Service take-rate}) \times (\text{Utilization factor})} .$$

Continuing our example, if the cable penetration is 65 percent in the area and 20 percent of the cable subscribers purchase cable modem service and we can estimate that no more than 80 percent of these would be connected simultaneously, then

$$\text{HPN} = \frac{54}{0.65 \times 0.20 \times 0.80} \approx 520 \text{ homes passed.}$$

If the service penetration increased to 25 percent, but all other variables stayed the same, then the HPN would be 105 homes lower. This emphasizes the difficulty of the estimating task, especially at the early stages of market development for these services.

Similar calculations need to be done for the other bandwidth-intensive services, such as telephony, to see which service has the most stringent HPN requirement.[8]

Two points need to be raised. The first is that there may be subscribers to, say, a cable modem service who are not subscribers to the entertainment TV service. Thus, counting data subscribers as a subset of cable subscribers may be an incorrect and market-limiting view. The second point is that unlike the connections we make over the telephone network, when we communicate over the cable upstream path, we are given "ownership" of a piece of bandwidth only for the instant when we are actually using it. This means that if a user's upstream message is only 100 kb long, five users can in fact communicate simultaneously at 500 kbps during a second. We have underestimated the system capacity by allocating that second to only one person in our estimate of $N_{simult,}$ above.

Future-Proofing System Capacity Without Overspending Today

The system planner is faced with a challenge: on the one hand, investment costs have to be minimized, which favors large nodes; on the other hand, future revenue opportunities can't be compromised, which favors small nodes. To make

8. After completing the return path traffic analysis, it is a good idea to check that the forward path bandwidth allocation is adequate (using the same method). In general, the systems being deployed utilize the forward spectrum more efficiently than the return band. For example, 64-QAM is used commonly in the downstream, while QPSK is more usual in the return. Because system designers realize that transmission robustness is at least as important in the return band as is spectral efficiency, they have opted for relatively simple modulation schemes for the upstream. Nonetheless, the adequacy of the capacity in the forward should not be assumed without double-checking.

matters more vexing, we don't yet have the ability to make accurate predictions of service penetrations and other important factors, as we have just seen. So what methods can we use to hedge the node-size decision and provide low-cost "insurance" that a decision made today will not be regretted tomorrow? The answers fall into five categories: node segmentation, architectures, block conversion, bandwidth efficiency, and expanded bandwidth.

Node segmentation

The internal design of many node products allows the operator to increase the number of return lasers as service needs require. This assumes, of course, that a sufficient number of unused fibers (sometimes called "dark" fibers) have been provisioned to the node in the original cabling. Figure 8.3 shows this schematically. While this is a very straightforward means for dealing with service expansion, it requires a large number of fibers. For example, if the 600-home nodes fed from a 24,000-home headend are subdivided into individual return quadrants, then the fiber count—for return only—goes from 40 to 160. As we have seen in Chapter Seven, the resultant increase in MTTR due to higher fiber count may cause the network availability to decrease.

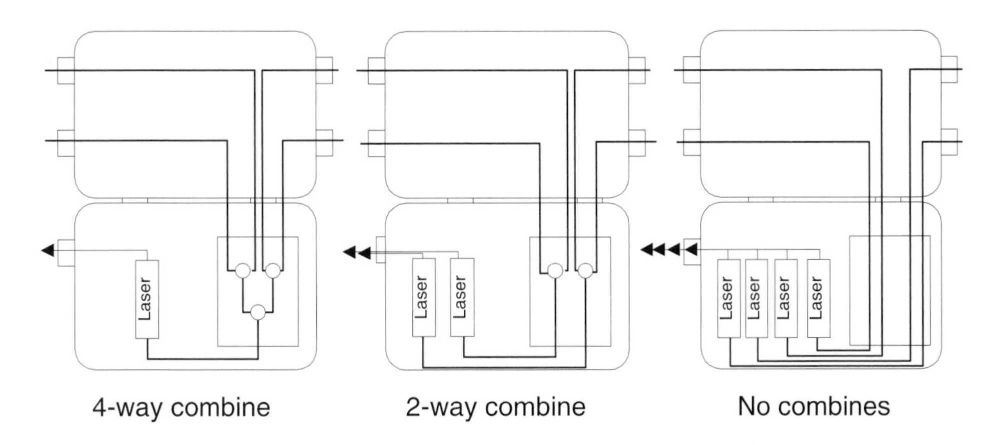

Figure 8.3 Node return segmentation

Architectures

A second method for future-proofing is to design today's node architecture in such a way that each node can be subdivided in the future at minimal cost. This is done by installing extra fibers to the node and laying out the RF distribution so that the first amplifiers after the node can be converted to nodes in the future (Figure 8.4). In this way, for example, an 800-home node can be replaced by four 200-home nodes as service penetration grows. Most manufacturers of distribution nodes make units capable of being converted to RF amplifiers and vice versa without resplicing the hardline cable. Notice in the figure that the original node has been removed in the future network. This means that the distribution must be designed originally with the subscribers in the initial node area being back-fed from the RF amplifiers. In many cases the coax connections from the original node are left in place to provide powering to the four sub-networks.

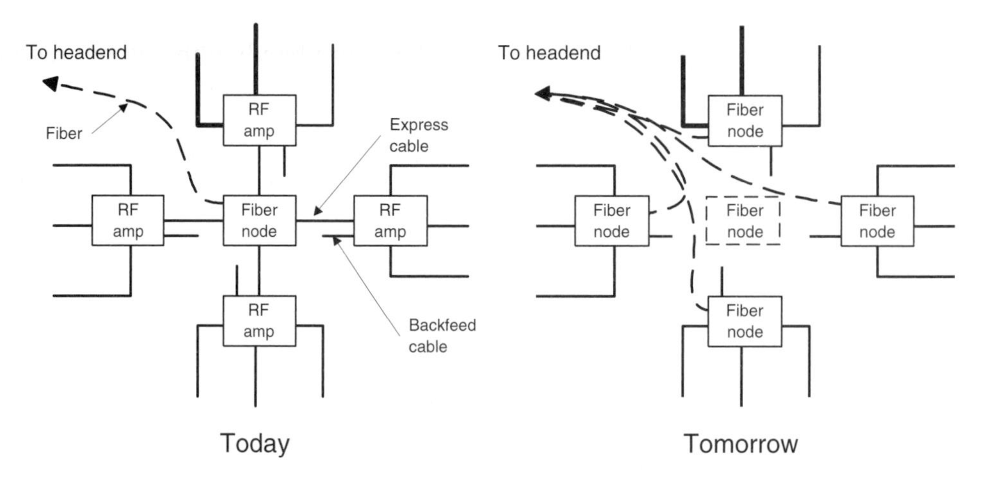

Figure 8.4 Node subdivision

A variant on this concept retains the original node and the coax connections (Figure 8.5). In this arrangement, the new nodes have forward fibers only, and the original node is used to collect the return signals from these nodes and to transmit them upstream over fiber. As we will see in the next section on frequency stacking, this does not necessarily compromise the return capacity, and it does reduce both the number of fibers back to the headend and the number of return lasers.

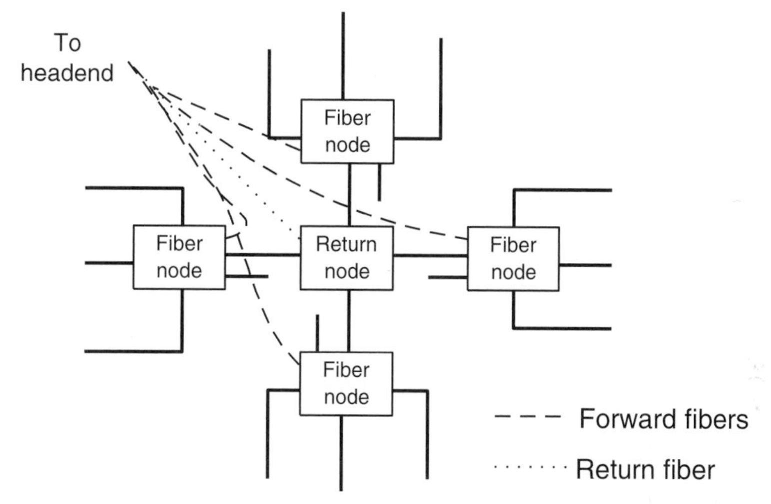

Figure 8.5 Return collection node

Block conversion

An alternative approach that increases the return bandwidth without reconfiguring the network or increasing the fiber count is to make changes within the node electronics to permit *block upconversion* or *frequency stacking*. In this scheme the four return legs feeding the node are not immediately combined. Instead they are individually shifted upward in frequency so that the four resulting bands do not overlap one another. The new signals are then combined and transmitted by a single laser to the headend.

A representative spectrum diagram is shown in Figure 8.6. The actual choice of frequency plan involves a trade-off between minimum guard band size versus filtering requirements for band separation and image rejection. Note that since the combined bandwidth is now much greater than the original 35 MHz, a somewhat higher performance laser is required, but this is not a major impediment. It is clear that isolation between the forward and return signals within the node becomes critical, since the two frequency bands now overlap.

After the optical signal is received at the headend, the upconverted blocks must be downconverted so that the signals are compatible with the application receivers. The up- and downconversion processes can add frequency offsets and phase noise to the received signals, which can potentially degrade BERs,

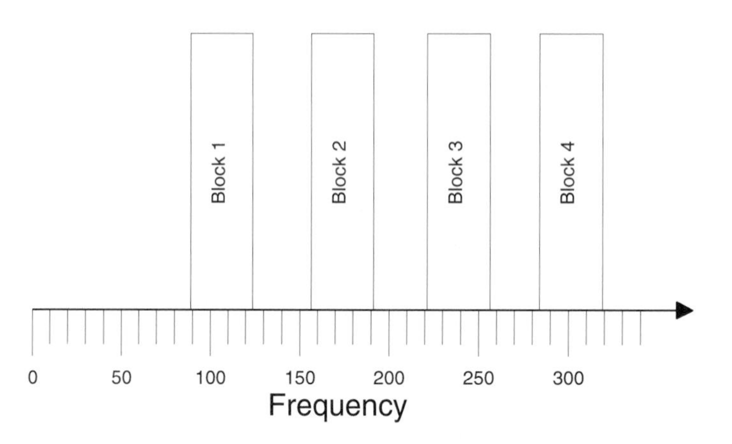

Figure 8.6 Block upconversion

increase latencies, or, worse, prevent signal capture altogether. The amount of these impairments depends on the specific design for the conversions. For example, conversion systems that carry a CW reference tone from the node can be designed to have essentially zero frequency error. In any case, a full system analysis is required to account for both the application requirements and any plant impairments.

An apparent advantage of the return collection scheme described in the previous section is that the upconversion electronics can occupy the entire volume of the former node station. In addition, since the forward signals do not pass through that station, concerns about forward-return crosstalk are eliminated.

Higher bandwidth efficiency

In the event that more communications capacity is needed, another alternative is to increase the spectral efficiency of the modulation. Per Table 8-2, a change from QPSK to 16-QAM, for instance, could permit twice as much data throughput. There are two immediate problems, however. The first is that the home terminal equipment would need to be changed, unless it had been built with the capability to change modulation scheme by remote command. The second is that the plant C/N requirement is tighter for the higher modulation type (by 6 dB, according to Figure 3.13).

Increased return spectrum

In the longer term, two alternatives could ease a return bandwidth bottleneck by actually increasing the size of the return band. The first is to obtain relief from the requirement (in North America) to carry the programming from channels 2 through 6 on their over-the-air frequencies. As can be seen from Table 8-3, if the downstream programming presently transmitted over the cable plant on channels 2–6 were relocated to frequencies above the FM band, the upper end of the return band could be extended to about 66 MHz (giving a 22-MHz guard band between return and forward). For a 5–40 MHz return system, this would be an increase of 16 MHz, or 46 percent. The effective increase is even greater since the low frequencies (5–15 MHz) tend to be noisy, hence less useful. There are clear institutional barriers to making such a reassignment. In addition, the multitude of forward path devices whose outputs are at channel 3 or 4, such as VCRs and set-top converters, and the TV sets that they feed would need isolation from return path emissions. Another hurdle is the conversion of all of the distribution equipment to the new band split. With equipment that employs plug-in diplex filters, this might not seem too difficult, but there are likely to be small subtleties in the equipment design that somehow carry the "signature" of the 40/52 bandsplit, such as flatness circuits and hardwired filters for snubbing forward signals from the return path. Last, if the operator has employed return equalizers of the "flat loss" type (Chapter Five), these would all have to be changed, which would be prohibitive.

Despite the impediments to revising forward channel assignments, however, the benefit of having such a great increase in clear return bandwidth may be sufficient to propel the cable industry to making the change.

Another—admittedly farther out—suggestion[9] is to put the return band above the forward band. This would relieve most constraints on the size of the return band. For an existing system, however, this would require separate return amplifiers and some type of coupling into the cable system, as described in the reference, because of the much higher RF attenuation at the high frequencies.

9. J. Chiddix, D. Gall, and G. Shimirak, "A Migration Strategy to High Capacity Return on HFC Networks," NCTA Technical Papers, pp 20–210 (1996).

Table 8-3 EIA channels in frequency order

Channel type	Channel designator	Carrier frequency (MHz)
Over-the-air	2	55.25
"	3	61.25
"	4	67.25
"	5	77.25
"	6	83.25
FM	Band	88 – 108
Cable TV	98	109.275
"	99	115.275
"	14	121.2625

Return Applications and Their Specific Requirements

In this section we will review each of the return service applications and discuss the ways in which they put particular requirements on the system. This is not meant to be a complete description of how each type of application equipment functions. Much of that information is now available on vendor Web sites.

Cable modems

The nature of cable modem service is very different from telephony, as is summarized in Table 8-4. Traffic characteristics for telephony are symmetrical and continuous, while data traffic can be highly asymmetrical and bursty. Typical of the latter are file downloads from the Internet, which require a few keystrokes upstream (entered at a person's typing speed) followed by a megabyte of information downstream. In fact, the rapid swing to Internet access via telephone modems caught the public telephone network with some capacity shortfalls. Keep in mind that the telephone system design assumed that handsets would be off-hook for relatively short conversations. Internet sessions tend to be lengthy, however, which means that a switch circuit must be devoted to an individual line for a longer period than had been planned. Cable modem systems—with their

LAN heritage—are naturally configured for bursty data and dynamically assignable capacity (Table 8-5).

Table 8-4 Characteristics distinguishing telephone service application from modem application

	Telephone service	Data service
Bandwidth/customer	Fixed	Variable
Traffic directionality	Symmetrical	Asymmetrical
Traffic flow	Continuous	Bursty
Latency tolerance	< 2 msec	< 10 msec

Since it is safe to assume that broadband communications will only increase in popularity and intensity, there is a need for bandwidth efficiency and bandwidth extendibility. There are a number of means for accomplishing these goals, as outlined earlier in this chapter.

Table 8-5 Characteristics distinguishing data delivery by telephone modems from cable modems

	Telephone data modem	Cable data modem
Bandwidth	Limited, fixed	Dynamically assigned
Connection	Dial-up	Always on-line
Standards	Well established	In development
Internet access	Unbalances network	Inherent
Security	Inherent in physical layer	In software, control and encryption

Standards

It is the intent of most cable operators to establish standard protocols and specifications for these systems so that their customers can purchase cable modems at retail stores and install them just as they are accustomed to do with telephone modems. At least three groups are developing cable modem system standards: a consortium of North American cable operators[10] called the Data Over Cable Service Interface Specification (DOCSIS) working group,[11] the

IEEE Computer Society's 802.14 Cable-TV Protocol Working Group,[12] and the Digital Audio-Visual Council (DAVIC).[13] All intend to submit standards to the International Telecommunication Union (ITU, formerly CCITT). The protocols will incorporate privacy safeguards, since the shared nature of the physical network in a cable system means that security is not inherently provided.

Equipment

A cable modem system consists of the major elements shown in Figure 8.7, which follows the DOCSIS nomenclature. Initially, most organizations are deploying modems that are external to the user's personal computer. This utilizes a well-established standard, such as IEEE 802.3 (twisted-pair Ethernet), Universal Serial Bus (USB),[14] or IEEE 1394 (FireWire)[15] between the modem and PC, thus reducing the difficulty of interfacing. Although an internal modem would be less expensive, it is a much larger challenge to ensure its compatibility with all makes and configurations of PCs. The modem transmits and receives via a direct connection into the cable system through the in-home cabling. At the headend, the modems communicate through a device generically called a Cable Modem Termination System (CMTS). The CMTS provides the needed upstream and downstream interfaces between the modulated carrier HFC system and a wide area network (WAN) that connects to routers, servers, and the Internet.

The cable modem is assigned an *Internet Protocol* (IP) address at the time of service start-up. Since the cable network provides an open pipeline to that IP

10. Multimedia Cable Network System Partners Ltd. (MCNS), which is a partnership of Comcast Cable Communications, Cox Communications, Tele-Communications Inc. and Time Warner Cable, along with MediaOne, Rogers Cablesystems, and Cable Television Laboratories (CableLabs). Arthur D. Little prepared the initial specification documents, under contract with MCNS.

11. Information and specifications are available at http://www.cablemodem.com.

12. http://802.14.org.

13. http://www.davic.org.

14. http://www.usb.org.

15. http://www.skipstone.com/info.html.

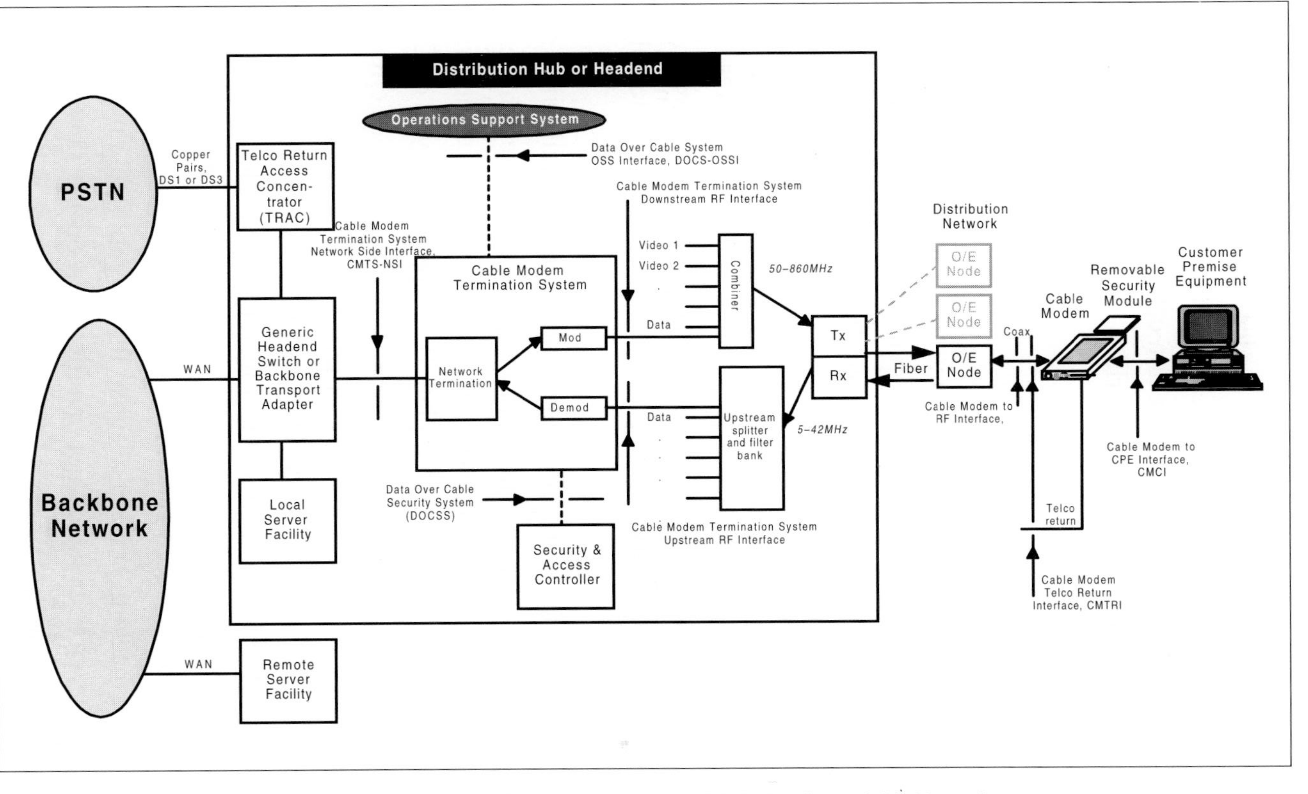

Figure 8.7 Data Over Cable System Diagram (by permission of MCNS Holdings, LP). Note that telephone return is included in the diagram.

address, the modem and PC can remain connected to the Internet at all times, if the subscriber desires. This puts relatively little burden on the capacity of the network, unlike the conventional telephone modem, which holds onto a switched connection for as long as it is connected. As a result, cable modems allow a home or small business PC to be used directly as a World Wide Web site that is accessible at any hour.

Telephony

One of the characteristics of developed countries is a well-established local telephone service. This implies that the only opportunity for new companies wanting to provide telephone service in those countries is to displace customers from the existing telco to their own service. The three most direct ways to achieve that displacement are to offer service that has either higher quality, more features, or lower costs. In the US, the most probable course is to offer equal service at lower cost, principally through bundling the phone service with other attractive offerings.[16] But what does it take to provide comparable telephone service quality?

First, the reliability of the service must meet the perceived availability standards set by the existing providers. As noted in the previous chapter, this establishes stringent requirements—ones that need to be incorporated in the system design criteria from the outset of any rebuild plan. Based on the analysis of Chapter Seven, that performance level should be achievable, with appropriate effort.

Second, the cable plant must be capable of powering the customer's handset independent of the local electric utility. This permits use of the phone system even during an electrical power outage and is a feature of most local telephony services. In the US, this is usually achieved by providing battery backup that is, in turn, backed up by a gas-powered generator when necessary.

This independent backup powering capability is required uniquely for wireline[17] telephony service. It has major ramifications on the design of the cable system and its equipment:

16. Offering PCS service, which we discuss later in this chapter, is perhaps an example of "more features," since its wireless operation provides new capabilities.

- The amount of power distributed through the system must be increased.
- Power delivery down the drop must be provided.
- Standby batteries in all AC power supplies must be provided.

We will discuss each of these powering issues after a brief overview of HFC telephony equipment.

Equipment for HFC telephony

In an HFC telephony system, the telephones in the home are connected by conventional twisted-wire pairs to a small box on the side of the house, generally called a *network interface unit* (NIU). The NIU puts the telephone calls onto RF carriers for transport over the cable system. At the headend, the carrier-based signals are demodulated by a *host digital terminal* (HDT). The HDT, in turn, communicates with a telephone network switch. Typically a single HDT can handle 672 simultaneous telephone calls.[18]

Power consumption

We will evaluate how much additional power is required for telephony service beyond that needed for operating the HFC distribution plant, itself.[19] In order to provide sufficient power for telephony, we need to estimate the power required during the peak telephone service hour. Telephony traffic levels are usually measured in units of hundred call seconds, abbreviated CCS (the first C is the Roman 100). If, during the peak hour, a phone is off-hook (in-use) for 30 minutes, then it represents (30 min) × (60 sec/min) / 100 = 18 CCS. The telephone industry has developed models for the usage statistics of populations of

17. We use the word "wireline" to distinguish phone service provided by wired connection from PCS and cellular, which use over-the-air links. This service is sometimes called "lifeline" service because of its ability to operate in emergencies that include power outages. This is somewhat confusing because the U.S. telephone companies use the term "lifeline" to refer specifically to billing plans with minimal outgoing calls included, that allow low-income subscribers to have access when needed, at low cost. PCS and cellular are often used as secondary communications systems, providing mobility, with a wireline connection to the home as primary.

18. This is DS-3 capacity in the North American hierarchy (see Table C1-2 in Appendix C1).

19. Recall that in Chapter Six we calculated the power requirement for a 500-home HFC node and distribution.

phones with an average of 9 CCS per phone line during the peak hour. This is summarized in Table 8-6, showing how many phones are ringing and how many off-hook for various total CCS levels.[20]

We need to make some assumptions about the nature of the cable telephony equipment. First, assume that there is a side-of-the-house box (NIU) that separates the telephony signals from the other services. The NIU connects twisted-wire-pair cabling inside the house to the coax plant. It also branches out multiple phone lines. We will assume that each NIU can provide two lines to the home, since it would probably be short-sighted to construct for only one from the outset. (The average for the US is 1.2 phone lines per residence.) Some small amount of power is consumed by the NIU for second lines that have not been activated. In addition, to minimize power usage, the telephony equipment should have a "sleep" mode that keeps only the ringer-detection circuitry alive when the handset is on-hook (not in use). Power use is highest when the phone is ringing.

Now we are ready to model the power consumption for a cable telephone system served by the 500-home node that we analyzed in Chapter Six.

Assumptions:

Penetration rate for cable = 70 percent

Penetration for telephone = 10 percent (of the cable subscribers)

Activated phone lines per NIU = 1.1

Peak usage = 9 CCS per activated line

Power consumption per line[21]

Ringing	7.0 W
Off hook	6.5 W
On hook	2.0 W (Sleep Mode)
Inactive line	0.5 W

20. Note, for example, that 5400 total CCS means that there are 600 phones, since the model assumes 9 CCS per phone line. Nine CCS means that at all times during the peak hour, one quarter of the phones are off-hook.

21. These consumption figures may be on the high side, but they show the impact of telephony powering and the need for power efficiency.

Table 8-6 Telephone service statistics

Total CCS	Ringing	Off-hook (in use)
400	3	20
700	4	30
1000	5	40
1500	6	56
2000	7	72
2500	8	89
3250	9	107
3900	10	125
4600	11	144
5400	12	167
6200	13	190
7100	14	215
8050	15	242
9000	16	269

Reprinted with permission from Bellcore TR-NWT-000057, "Functional Criteria for Digital Loop Carrier Systems," Issue 2 (January 1993), Morristown, NJ.

The number of telephone subscribers is 500 × 70% × 10% = 35, which (at 1.1 active and 2 total lines per NIU) makes the number of active telephone lines 39 and the number of inactive lines 31. Hence there are a total of 39 × 9 = 350 CCS during the peak hour for the node. Per Table 8-3, this means that 3 phones will be ringing and 20 phones will be off-hook during the peak usage period. This allows us to calculate the peak hour power:

$$\begin{aligned} P_{telephone} &= N_{ring} \times P_{ring} + N_{off} \times P_{off} + N_{on} \times P_{on} + N_{idle} \times P_{idle} \\ &= 3 \times 7.0 + 20 \times 6.5 + [39-(3+20)] \times 2.0 + 31 \times 0.5 \approx 200\ \text{W}. \end{aligned}$$

In Chapter Six we estimated the HFC distribution power at 1050 W for a 500-home node. In this telephony model, peak hour telephony adds almost 20 percent to the distribution power requirement.[22] Since the sleep mode consumption for each line is approximately equal to the power usage of the RF equipment on a per-home-passed basis, the power budget is very sensitive to the service penetration assumption. At twice the penetration (20%) and twice the usage (18 CCS/line), telephony adds 50 percent to the distribution power requirement. It is obvious that the telephony equipment needs to be designed with energy efficiency—in all modes—as a primary consideration.

We've determined how much power must be delivered to the telephony subscribers. Now we have to figure out how to get it to their homes. There are three main considerations:

- To centralize power, if possible, because each of the power supplies needs to be backed up with batteries or generators, which need maintenance attention, and because finding sites for these units is difficult.
- To stay within the current ratings of the power supplies and the elements in the current distribution network, such as amplifiers and taps.
- To minimize the resistive losses in the distribution cable.

All of these considerations argue for raising the distribution voltage, and, indeed, a number of plants are now being designed and equipped for 90 VAC operation. In addition, some designs are utilizing *express power cable* to distribute power in parallel to the RF distribution (Figure 6.2). Using cable that is optimized for carrying current will increase the distribution efficiency.

A study done by our company shows the use of 90 VAC and express powering for a cable phone service with 25 percent penetration rate (Table 8-7). This work demonstrates the advantage of 90 VAC, which totally eliminates the need for adding express power cables in this case.

22. Keep in mind that this 20% applies to the plant capacity, hence its capital cost, but not to its operating cost. The 20% additional power is during the peak usage period, whereas the distribution plant load is continuous for 24 hours per day.

Table 8-7 Voltage and power cabling study

Voltage	**60 VAC**				**90 VAC**			
Subscriber distribution	Uniformly distributed		Clustered at extremities		Uniformly distributed		Clustered at extremities	
CCS rating	9	36	9	36	9	36	9	36
Number of 15 amp supplies	4	4	4	4	4	4	4	4
Express power cable	.860" to 1st active	.860" to 1st active	.860" to 1st active	1.125" to 2nd active	None	None	None	None
Express feeder cable (.715")	No change	No change	No change	No change	No change	No change	No change	Use .860"

A final power distribution consideration is the power factor of the distribution and telephony equipment. All of the foregoing discussion has made an unrealistic assumption that all NIU equipment has a 1.0 power factor, which would mean that the AC power required would be the product of voltage and current. In reality, capacitive circuits tend to cause the current to be out-of-phase with respect to the voltage, which increases the amount of current required for a given power. That increase in effective current needs to be included in the design calculations. Thus the power factor of the NIU equipment becomes an important consideration.

Drop powering

Having delivered the requisite power up to the tap, we now need to bring it down to the NIU at the side of the house. Two approaches are being used: (a) applying the AC power directly to the coax drop cable and (b) using so-called *siamese cable* consisting of coax for delivering RF and twisted-wire pair for power. Siamese cable has a figure-eight cross section, so that the two independent conductor systems are joined by a web into a single mechanical package. The advantage of using the existing coax is, of course, the great cost saving. There are some residual concerns about corrosion, which cause some operators to favor the Siamese approach. Needless to say, for the coax powering option, care must be taken in the NIU design to ensure that the AC voltage does not pass through to the in-home cabling. And in either case there needs to be a current-limiter in the tap to keep excessive currents from being passed to the home. This is needed both for safety reasons and for keeping the distribution plant from being pulled down by an NIU wiring fault.

The internal construction of a *power-passing tap* is also very different. The directional couplers that tap off the RF signals are not capable of passing current of any significance, so a separate circuit and connections must be made. Also in order to provide uninterrupted service at locations downstream from the tap, the unit must have an internal make-before-break bypass path that is activated when the tap faceplate is removed.

Battery backup

In Chapter Six we discussed the impacts of outages of the electric power utility. If the cable operator is providing telephony service capable of operating during such an outage, standby power is required at the AC power supply locations. As pointed out in the earlier chapter, four-hour battery capacity is a help, but it is essentially impractical to expect to span all utility power outages with batteries. In a small number of installations, AC generators have been deployed for that reason. The generators, generally fueled by natural gas, add to the problem of finding sites for the power supplies, of course.

Modem telephony and videophone

After considering the difficulties and expense associated with providing power for telephony, many operators have decided to ignore this application altogether. There are two alternative approaches that do not require the powering infrastructure, but still maintain access to a significant portion of the telephony revenue opportunity. We will discuss PCS over cable in the next section. In this section we discuss ways to provide second-line service and videoconferencing. By "second-line" we mean that the home will have a primary phone line provided conventionally by the local telco. All other phone lines—for fax machines, for other family members, or for a home office or business—don't actually require powering that will span an outage of the local power utility.

With the burden of uninterruptible powering removed, the infrastructure for telephony over cable is not very different from that of cable modems. The primary difference is that the telephony modems and the headend units (HDTequivalents) need to be designed for symmetrical traffic and for defined bandwidth allocations so that there is no appreciable latency (delay) in communications in either direction.

This type of modem can be combined with the abundant two-way bandwidth available in HFC networks to offer high-quality but affordable videophone or videoconferencing capability. Videophone *codecs* (combined digital compression encoder/decoders and modems) running at 384 kbps are becoming available for these applications, providing very good resolution and motion reproduction. The ITU has developed standards (H.323) covering this type of video telephony.

Videoconferencing customers require highly reliable transmissions, hence the operator needs to be able to provide that level of quality. Operators should not view this, or any other second-line opportunity, as being "second class." Only the powering requirements have been eased.

Personal communication services (PCS)

PCS telephony is a low-power, microwave-carrier version of the traditional cellular services. It employs much smaller cells than "cellular" (approximately 1 mile radius vs. 5 mile), it operates at higher frequency (1.85–2.00 GHz vs. 800–900 MHz), and it relies inherently on digital transmission technologies, such as CDMA and TDMA. PCS offers the convenience of wireless telephony, but reduces some of the disadvantage of battery dependence because handset power consumption is reduced by the small cell size.

Being a late arrival on the communications scene, the PCS operator needs to deploy systems quickly. Since the operator has invested heavily to acquire operating licenses, cash flow needs to managed carefully. In many cases, both cash flow and speed-to-market can be aided by connecting PCS antennas to the cable plant for these reasons:[23]

- Antenna siting problems can be reduced because the antennas become part of the existing cable system.

23. Here we are assuming that the cable operator and the PCS operator are not the same company, which is not always the case.

- Coverage can be increased by simply interconnecting adjacent antennas at the headend to make larger cells.

For either of these applications, the cable system is used primarily for bidirectional trunking. The PCS signals are downconverted to RF frequencies in the cable forward and return bands for transport between the PCS switch center and antennas that are mounted on the cable strand. As shown in Figure 8.8, strand-mounted *cable microcell integrators* (CMI) serve as PCS antennas and as frequency translators between PCS and cable bands. They communicate over the cable network with *headend interface converters* (HIC),[24] which do the opposite frequency conversions and connect with a *base transceiver subsystem* (BTS). The BTS, sometimes referred to as a *base station*, hands off and accepts calls from a *mobile switching center* (MSC), which provides the appropriate interface between the PCS system and other local and long-distance telephone services.

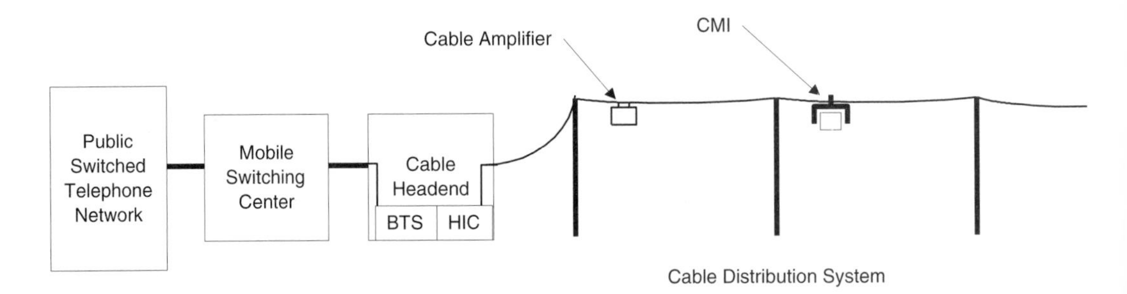

Figure 8.8 PCS over cable

It is important to understand that the PCS-over-cable system does not communicate with the subscribers' homes via the drop cable, since the cable system is used only for trunking of aggregated PCS signals. The link to the subscriber is over-the-air in the 1.85–2.00 GHz band. Thus, PCS-over-cable is the one interactive application that *is* compatible with the return blocking filters discussed in

24. Initially the term *Remote Antenna Driver* (RAD) was used in referring to the CMI, and *Remote Antenna Signal Processor* (RASP) was used for HIC.

Chapter Five. We will see an implication of this later in this chapter when we discuss spectrum allocation.

Headend Equipment Requirements

In the headend there will be one fiberoptic receiver for each node (with no fiber redundancy).[25] If there are N different services running through the homes served by that node, then the RF output will be split N ways to provide signals for each of the application receivers (Figure 8.9). For most applications, these cables will connect directly to the application receiver inputs. For other applications, such as polled set-tops and status monitoring, it may be acceptable to combine several cables into one port of the application receiver. Furthermore, in the initial stages of market development, it may be possible to economize by combining even the high-revenue applications, like modems, until traffic builds.

RF amplifiers may be needed for extensive signal fan-outs when there are multiple applications. Also, some ability to adjust levels and to equalize will be useful since the headend cables may be lengthy. TCI Engineering has prepared an overview that provides a convenient framework for considering headend cabling and equipment.[26] Our treatment of this subject is in Chapter Eleven.

Spectrum Allocation

In Chapter Four we discussed that there were two troublesome types of noise and interference: diffuse noise that is primarily at the low end of the return band and discrete interference that can be anywhere, but often shows up at predictable locations (such as amateur radio frequencies). Now that we have described the characteristics of the return applications, what suggestions can be made about allocating the available spectrum?

First, the discrete interferers should either be controlled or their frequencies avoided. We have learned from early experience with upstream applications not

25. In Chapter Nine we discuss why it is not a good idea to combine optical returns from multiple nodes into a single receiver.

26. O.J. Sniezko, "Multi-Layer Headend Combining Network Design for Broadcast, Local and Targeted Services," NCTA Technical Papers, pp 300-315 (1997).

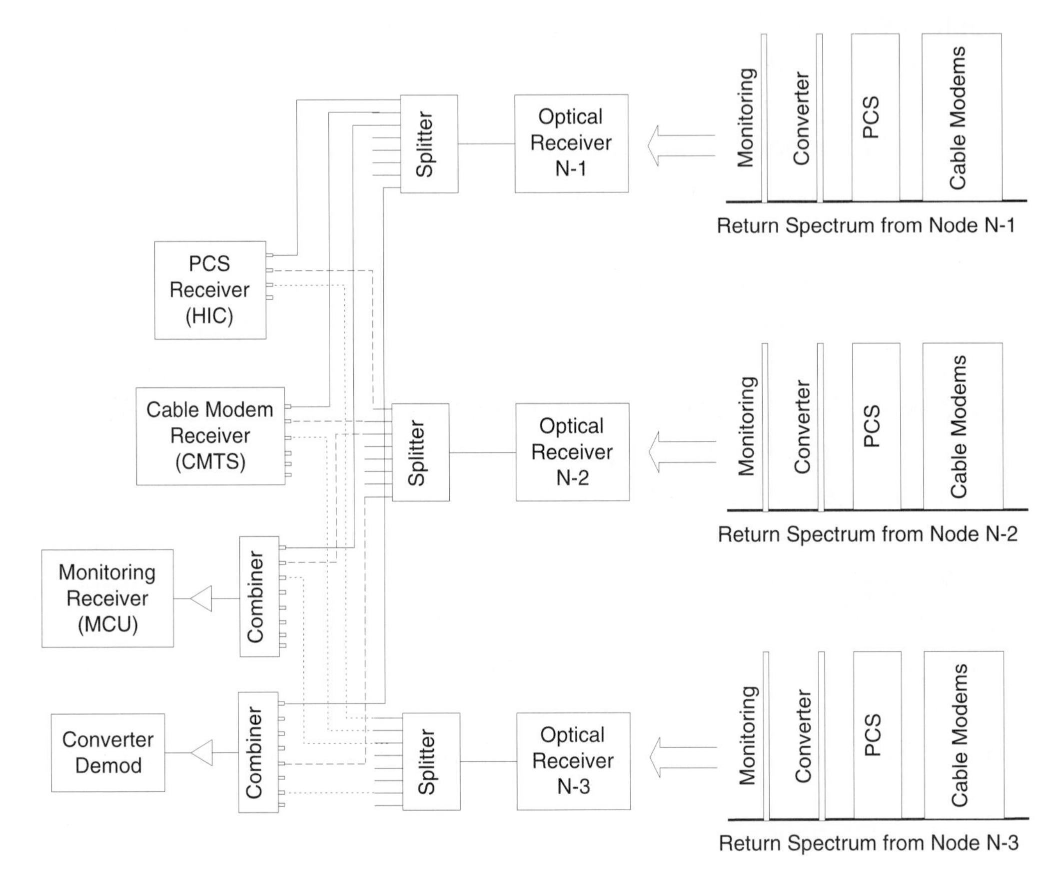

Figure 8.9 Return path interconnection in the headend

to combine a number of users onto one wide-band signal. This is because the entire channel can be compromised by a single narrow interferer. It is much better to carry the data on a multiplicity of subcarriers within that same allocation of bandwidth, as is commonly done now. This permits unimpeded use of most of the band, even in the presence of an interfering tone. The application controller needs to monitor the quality of the communication sub-channels to determine which sub-carriers should be used.

There are two observations relevant to operating in the presence of the diffuse low-frequency noise. The first is that some signals, such as the polled set-top returns, may be able to work in that environment, due to their lower C/N requirement. The second is a deployment idea being implemented by Cox Communications in San Diego, CA: to use the lower portion of the band for PCS. While one is disturbed at first by the thought of operating any kind of telephony in the "dirty" part of the band, it is actually a very good fit. Recall that PCS uses the cable distribution only for trunking and does not carry signals directly to or from the home. This means that 5–20 MHz blocking filters can be put into all of the taps. As a result, the hardline cable should be clean in that band and ideally suited for PCS trunking. Interactive services from the home can be put above 20 MHz. For systems with a large installed base of two-way set-tops that operate below 20 MHz, window filters may be effective.

The most demanding applications are the ones (like wireline or PCS voice telephony) that cannot tolerate the delays inherent in retried transmissions. This means that they should be allocated the cleanest portion of the spectrum.[27]

Summing-Up...

- A straightforward method exists for estimating capacity requirements.
- HFC architectures and equipment provide several options for upgrading capacity as markets build.
- The cable modem application is well suited to the capabilities of an HFC network.
- Wireline telephony is more difficult, but it can be done.
- PCS is a more natural fit for HFC than is wireline.

27. Alternatively, they should be allocated extra power, if the constant-power-per-Hz method is not being used (as will be explained in Chapter Ten).

CHAPTER 9

Fiberoptic Links for the Return System

In this chapter we will discuss how to predict the performance of fiberoptic links in the return system. We will describe the different types of return lasers and explain the differences in their performance. This will establish the basis for choosing a return laser.

Types of Return Lasers

All lasers consist of three basic elements: a laser cavity, an optical pump, and a mirror system (Figure 9.1). The laser cavity is created in an optical material with these properties:

- Electrons in the material can be excited into an upper energy state that is semi-stable.
- An optical waveguide can be formed in it.
- Light of the desired wavelength is not absorbed by it.

A laser cavity is formed by the optical waveguide and two end mirrors. With the aid of some type of "optical pump," electrons in the waveguide region are excited. The key phenomenon that makes lasers possible is that when the light bounces back and forth in the laser cavity, it can make these electrons fall back to their ground state energy, thereby generating more light that is precisely in-phase with the stimulating light. If there are enough excited electrons—or, as we say, if the laser is operated above threshold—this increases the light intensity. The word *laser* is an acronym for Light Amplification by Stimulated Emission of Radiation.

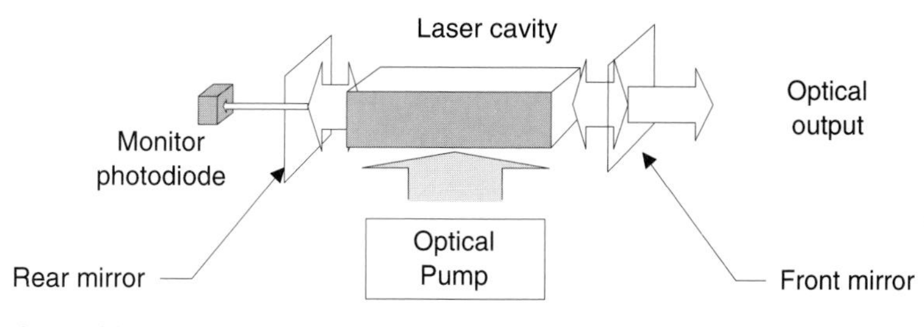

Figure 9.1 Laser block diagram

In order for the device to be useful, of course, some of the light must be allowed to pass through the front mirror. In the lasers used in telecommunications, this output light is lensed into a singlemode optical fiber.

All return path lasers are *semiconductor lasers*, which means that the lasing material is a semiconducting material—usually an alloy of the elements gallium, indium, arsenic, and phosphorous grown on a substrate of indium phosphide—and the optical pumping is done by passing an electronic current transversely through the cavity. The end mirrors are formed by carefully cleaving the semiconductor wafer and adding an optical coating. The laser module includes a photodiode that collects a controlled amount of light leaking through the rear facet mirror. This *monitor photodiode* is used in a closed loop optical power control circuit. The monitor photodiode, laser, lensing system, and fiber attachment are contained in an electronic package that becomes a component of the return path transmitter module in a fiberoptic node.

Two additional pieces of laser technology will help us understand return path performance. As we said, the light bounces back and forth between the end facet mirrors. For given cavity dimensions and optical characteristics, there will be a set of evenly spaced optical wavelengths that can stay in the cavity without being attenuated. This is for the same reason that an organ pipe carries only a specific set of tones. In addition, only a narrow band of optical wavelengths can stimulate the excited electrons in the laser medium to emit radiation. This gives rise to the two phenomena shown in Figure 9.2: the comb of allowable laser wavelength modes and the semiconductor *gain curve*. Light can propagate and be emitted at all of the allowable wavelengths, but amplification is greatest at the center of the gain curve.

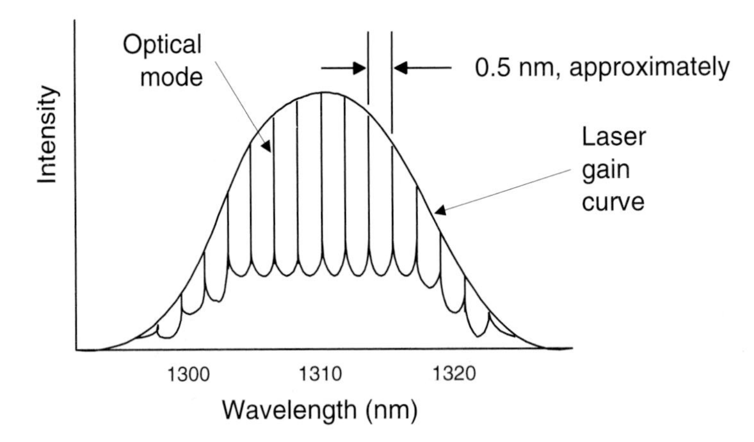

Figure 9.2 Semiconductor laser characteristics

The laser wavelength comb

The wavelength spacing shown in Figure 9.2 is a result of interferences between the optical waves moving forward and back in the laser cavity. The interference is constructive when the round trip for a wave is an integer number of full wavelengths. The wavelength in any medium is the vacuum wavelength λ_0 *divided by the material's optical refractive index n. In this case* λ_0 *is 1310 nm, and n is approximately 3. If the cavity length is 0.5 mm, then it will take the following number of wavelengths to make a round trip in the laser cavity:*

$$N = (2 \times 0.5 \text{ mm}) / (1310 \text{ nm} / 3) = 2 \times 5 \times 10^{-4} / (1.3 \times 10^{-6} / 3) \approx 2300.$$

A slightly longer wavelength $\lambda' = \lambda_0 + \Delta\lambda$ *could make the round trip in one less period. That would make the next higher wavelength in the comb approximately (2300 ÷ 2299) times as long. Thus* $\Delta\lambda = (1/2299) \times \lambda_0 = 1310/2299 \text{ nm} \approx 0.5 \text{ nm}$*. By the same analysis we can show the general formula*

$$\Delta\lambda = \lambda_0^2/(2nL),$$

where L is the length of the laser cavity and n is the index of the cavity material. From this we can see that shorter-cavity lasers will have more widely spaced modes.

This describes the operation of a *Fabry-Perot* (FP) laser, whose optical output is characterized by many wavelength modes grouped around a central wavelength. When we speak of a 1310 nm FP laser, we are referring to its approximate center wavelength.

By adding more structure during fabrication of the laser, it is possible to suppress all modes other than the center wavelength. This is done by creating a washboard-like grating along the length of the cavity (Figure 9.3). It is not literally a mechanical corrugation in the laser structure; rather, it is a periodic variation in the optical properties of the medium just below the waveguide. The grating acts as an optical filter that reinforces the desired wavelength and diminishes all others. Because the grating extends over the length of the cavity, this is called a *distributed feedback* (DFB) laser. The grating is very effective: a typical DFB laser will have side modes suppressed by 50 dB or more.

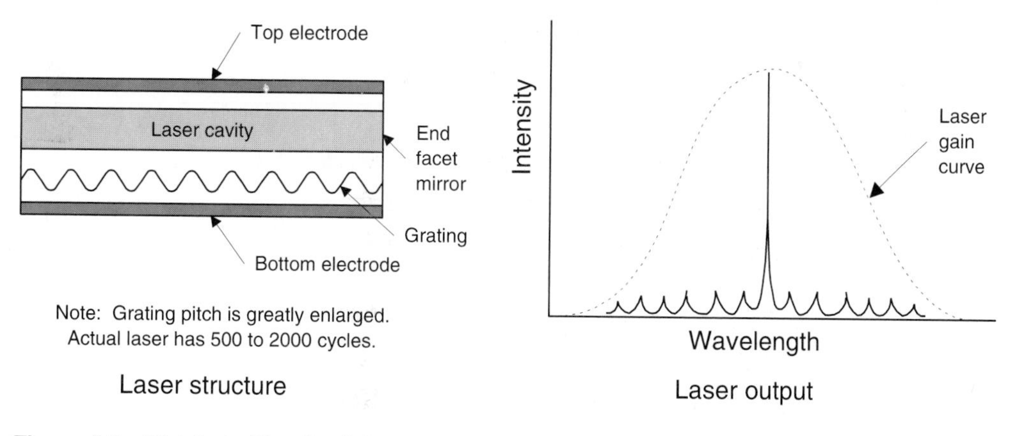

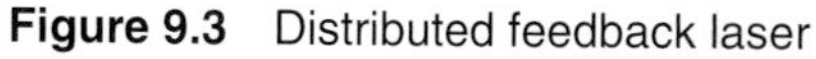
Figure 9.3 Distributed feedback laser

Either type of semiconductor laser can be *directly modulated*, which means that the RF signal is applied to the same electrodes as the bias current. This is accomplished with a bias-T circuit that inductively couples the DC bias current and capacitively couples the RF.

We will need one more definition, *optical modulation index* (OMI), which is defined using the quantities shown in Figure 9.4 (often referred to as the *L-I curve* of the laser).[1] The quantity ΔI (“delta I”) is the amplitude of an RF current applied to the laser, while ΔL is the resultant change in optical output power.

$$OMI = \Delta L \div (L_0 - L_{thresh}) \approx \Delta L / L_0 \quad \text{[optical definition]}$$
$$OMI = \Delta I \div (I_{bias} - I_{thresh}) \quad \text{[electrical definition]}$$

Either way of measuring OMI will give the same result since the laser is being operated in the linear region.[2] The OMI is an indicator of how much RF power is being applied to the laser.

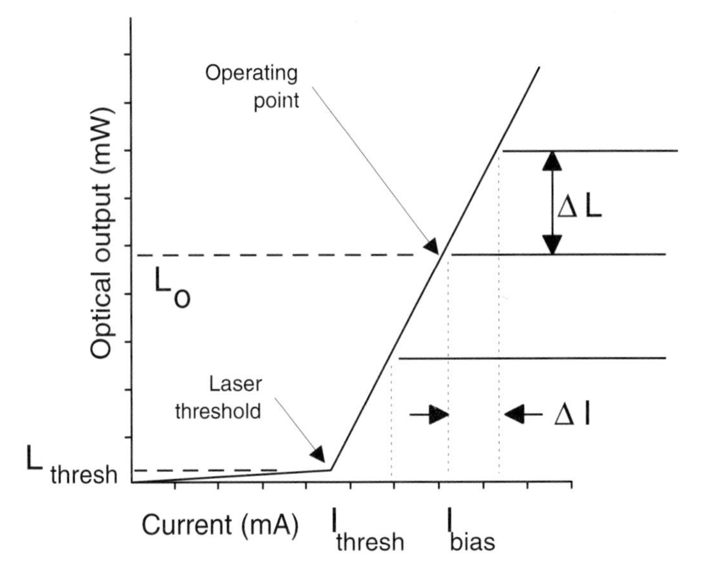

Figure 9.4 Optical modulation index

Fiberoptic Link Performance

The reason we need to devote more than two full chapters to fiberoptic links is that the performance of these links is the major determinant of how well the

1. In the L-I curve we can see that as the electrical bias current is raised initially from zero, the semiconductor diode emits a very faint optical output. Above threshold, it becomes a laser and emits strongly.

2. Note that the optical power at threshold is so small that it can be left out, but the threshold *current* is significant, hence must be included in the electrical calculation.

overall return system functions (provided that ingress is under control). Three characteristics are critical:

- ***Noise*** We have already shown in Chapter Four how the noise performance of the fiberoptic portion of the return path dominates the calculation of C/N performance for the entire system. To a degree, this is true of the forward path as well.
- ***Clipping*** Laser clipping sets the upper limit on the amount of RF power that can be passed through the return system. In Chapter Four we noted that when a laser is driven below its threshold, the optical output will be turned off (Figure 4.7). This results in a negatively clipped RF signal at the receiver. In the next chapter we will discuss the impact of this power limitation on return plant design. Unlike the forward path, other distortions caused by nonlinearities in the lasers do *not* appear to be a limiting problem for digital return signals
- ***Spurious emissions*** Commonly abbreviated as *spurs*, these show up as intermittent spikes in the return spectrum. They are generated by instabilities in the laser cavity caused by laser light that has been reflected back from the fiber.

C/N in return path fiberoptic links

Four types of noise will impact the C/N of the fiberoptic link (Figure 9.5).

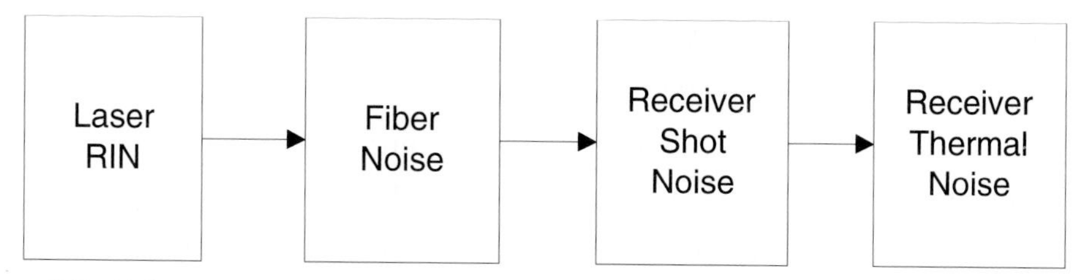

Figure 9.5 Noise sources in the fiberoptic link

We will discuss each of these in some detail to give the reader an understanding of the phenomena and of their relative importance in different lasers and in different applications. For the reader who needs the answer in a hurry, however, Table 9-1 indicates the areas for concern in each return laser situation. The signal impairments are ranked by the extent of their impact on the C/N

("Large" means highly adverse impact on C/N). Keep in mind that there are remedies for most of these concerns, as we will discuss later. Also note that a "large" threat of impairment does not imply that the laser is unusable in the application, merely that this is the main area for concern in that case.

Table 9-1 Principal impairments to C/N in return links

Laser Type	Distributed Feedback (DFB)				Fabry-Perot (FP)			
Isolator	with isolator		w/o isolator		with isolator		w/o isolator	
Fiber length	Short	Long	Short	Long	Short	Long	Short	Long
RIN	M	S	M	S	L	M	L	M
Fiber:								
MPN	—	—	—	—	M	L	M	L
Spurs	S	S	M	L	—	—	M	L
IIN	S	M	S	M	S	S	S	S
Shot	M	M	M	M	S	M	S	M
Thermal	S	M	S	M	S	L	S	L

Key: S = Small degree of impairment, M = Medium, L = Large; — = not significant. "Short" fiber links are 0–5 dB; "Long" ones are 5–10 dB.

Laser noise (RIN)

The first noise type, *relative intensity noise* (RIN), is an inherent property of the laser, resulting from instabilities within the laser cavity. Since this noise is generated in the laser source, its intensity is attenuated in fiber at the same rate as the signal, so the C/N due to laser RIN is independent of fiber length. For short links, RIN is the principal noise source (other than spurs and MPN, which will be discussed later). As fiber lengths are increased, other noise contributions become more significant.

Fiber noise

Recall the warning in Chapter Four that lasers must always be evaluated in links actually containing optical fiber, rather than with optical attenuator devices. This is because several noise-inducing phenomena take place inside the fiber. We will discuss each in detail later in this chapter. For DFB lasers used in the forward path, the predominant type of fiber noise is *interferometric intensity noise* (IIN), which is caused by double reflections of the light within the fiber. The performance of return path lasers is less affected by IIN, but more so by spurious emissions and—in the case of FP lasers—by *mode partition noise* (MPN), which is a result of fiber dispersion acting on the multi-wavelength light from the FP.

We are including spurious emissions in this group because spurs result from interactions between fiber reflections and the laser. Unlike the other noise phenomena, however, spurs are impulsive. They show up as random spikes of short duration, so to observe them on a spectrum analyzer one uses little or no averaging and sets the analyzer to peak-hold for 30 seconds of a 5 to 40 MHz sweep. Alternatively, they can be measured in the time domain by setting the spectrum analyzer at zero span and observing for 20 to 40 seconds. Spurs can be a problem with either type of laser.

Mode partition noise (MPN)

FP lasers are beset by a fiber noise mechanism called mode partition noise, which arises because fiber is a dispersive medium. Essentially all of the optical fiber used in cable TV is designed to have a *zero dispersion point* at 1310 nm, which means that the minimum propagation velocity for an optical wave is at a wavelength of 1310 nm in the fiber.[3] In the case of an FP laser, with its many wavelengths, the dispersion cannot be zero for all of the wavelengths, which means that the different wavelengths making up the FP output will travel at different speeds.

At any instant, the constant optical power emitted from an FP laser will be composed of differing amounts of power in each of the various wavelengths.

3. The propagation velocity is wavelength-dependent due to two phenomena. The optical velocity of the material in the center of the fiber is lower than that in the outer glass cladding, so short wavelengths—for which more of the wave energy is in the central glass core—go slower in the material. On the other hand, the waveguide structure of the glass causes longer wavelengths to go slower because more of the long wavelength's energy is near the waveguide containment. These two effects can be balanced for just one wavelength. Nearly all of the fiber currently deployed is designed to be balanced at 1310 nm.

Consider what would happen if a single RF tone were applied to the laser. The light arriving at any instant would be a mixture of light arriving "on-time" from the central wavelength plus "early" light from a "faster" wavelength and "late" light from a "slower" wavelength. Because the intensity of each of these different signals depends on the random population statistics within the laser (that is, what portion of the total optical power is in each of the wavelength modes at each instant in time), the intensity of the combined signal will have some degree of random variation on top of the RF tone. This is MPN.

Following the single RF tone example a bit further, it is easy to see that MPN will be increasingly troublesome as the difference in travel time between fast and slow wavelengths becomes a larger and larger fraction of the RF signal period. This can happen either when the time differences are large or when the RF period is small. Hence MPN increases with fiber length (because the time displacement due to dispersion is proportional to fiber length) and with higher RF frequencies.[4]

Spurious emissions

Microscopic random imperfections in the glass fiber, which act like mirrors for the light, can cause some of the laser light to scatter back into the laser.[5] This reflected light can induce the laser to change operating modes. The wavelengths of the different modes are nearly identical—differing by less than the laser linewidth. This sets up a scenario in which the photodetector can see two different signals that are uncorrelated in time coming from the laser. The first is a direct signal sent out in a higher-order mode. The second is a signal sent out earlier in

4. H. Blauvelt, I. Ury, P.C. Chen and T.R. Chen, "Return Path Lasers for High Capacity Hybrid Fiber Coax Networks," NCTA Technical Papers, pp 35–40 (1995).

5. The reflections, often referred to as *backscatter*, are a result of what is called *Rayleigh scattering.* This is the same phenomenon that causes sunlight to scatter off particles in the earth's atmosphere, resulting in the blue sky in daytime and the red-orange sky at sunrise and sunset. When discussing light scattering within optical fiber, however, it is important to remember two facts. First, very little of the light gets scattered, otherwise the fiber attenuation would not be as low as it is. Second, only a very small amount of the light that <u>is</u> scattered will be directed back toward the laser and stay within the fiber waveguide. Thus we are discussing effects due to exceedingly small intensities of light. We see these effects only because the laser light is coherent, so optical interference can occur.

the original mode that has been doubly reflected in the fiber (that is, a signal delayed by a random amount of time due to the extra path length of an internal round trip).

The detector sees two independent signals, each with a spectral width corresponding to the laser linewidth (which may be on the order of 10 MHz[6]). The photodetector is a square-law device, converting incoming power into current (square-root of power), hence it is inherently a non-linear device. While this is very useful, it has the unfortunate side effect of making the detector a very good frequency mixer, as well. Thus the two optical signals mix in the detector and produce random (that is, noise-like) f_1+f_2 and f_1-f_2 intensity beats. These beats will fall throughout the RF spectrum that lies below twice the laser linewidth (below 20 MHz, approximately). This will impact the low end of the return band, but some of the bursts are high enough in energy to produce significant harmonics, so it is not uncommon to see spurs throughout the rest of the return band, as well.

The higher-order laser modes that cause these spurious emissions last for very short times, thus the beats appear as impulsive bursts. As such, the spurs interfere with digital signals in somewhat unpredictable ways. High-order modulations, including 16-QAM, are more susceptible to spurious interference than QPSK. This means that spurs need to be minimized if reliable communications are to be carried over the return path.

Spurs are a concern for both DFB and FP lasers. In the case of FPs, however, the spurs are somewhat less noticeable because the background of MPN is so large. In a sense the FP is always operating in many modes, so these cavity disturbances due to backscatter don't appear to be as disruptive. The claim-to-fame for DFBs, on the other hand, is a very low noise level (low RIN and no

6. It is common to refer to the linewidth in frequency units, even though that may seem odd at first. Since frequency and wavelength are inversely related by the speed of the light ($f = v/\lambda$), there is a straightforward correspondence. In linewidth units, a laser line is on the order of 10^{-14} microns, which is almost incomprehensibly small. In fact, one cannot measure such "dimensions" with an optical spectrum analyzer or any other direct method. Rather, the linewidth is measured by homodyning (beating the optical signal with itself) to transform the width down to "baseband," hence the linewidth is actually measured as a frequency on an RF spectrum analyzer.

MPN) due to their single cavity mode operation, and this can be thrown into chaos by backscatter. The cure in either case is to add an optical isolator.[7] The degree of isolation required for DFBs is higher than that for FPs.

Demonstrating spurious emissions

The mechanism that produces spurious emissions can be understood from the following series of experiments.

Set-up #1: An unisolated laser feeds directly into an optical splitter. One leg of the splitter feeds 5 dB of optical fiber; the other leg feeds an optical attenuator of the same loss. Identical detectors terminate each leg.
Observation #1: Spurs are seen in the fiber leg, but very few in the attenuator leg.

Set-up #2: A high-quality isolator is inserted between the laser and the splitter. No other changes are made to the set-up.
Observation #2: Spurs disappear completely from both legs.

Set-up #3: The attenuator is replaced by an equivalent length of fiber, and the isolator is moved to the beginning of that leg (that is, after the splitter).
Observation #3: Spurs are observed on both legs, in equal amounts.

Discussion of the observations: Spurs are created by mixing between doubly reflected light in the fiber and short-duration modes that are produced in the laser as a result of single reflections (backscatter) from fiber. In #1 the laser is disturbed by the fiber backscatter (as evidenced by the fiber leg), but there are essentially no double reflections in the attenuator leg, hence no multiple signals at the detector and, accordingly, very few spurs. In #2 the backscattering is prevented from entering the laser by the isolator, hence there are no laser mode changes and no spurs. In #3, backscatters from the unisolated fiber can enter the laser, thus inducing higher-order laser modes. These modes mix with the double reflections produced in each fiber, causing spurs at both detectors. Since the isolator does not prevent reflected light from entering the laser, it is ineffective in this placement.

Interferometric intensity noise (IIN)

IIN is generated in the photodetector as a result of interactions between the direct signal and signals whose path has included multiple reflections within the

7. An optical isolator performs the same function in the optical domain as a diode does in the electrical. All optical isolators are based on the magneto-optical *Faraday effect*, which allows a magnetic field to change the polarization of light. Like all other magnetic effects it has a "handedness," so the polarization change is different for light traveling in the two directions. When crossed polarizers are applied appropriately, very high directivity (>35 dB) can be achieved. The devices can be made small and rugged, but tend to add significantly to the laser module cost.

fiber. At any point in the fiber, an imperfection may cause some of the light to bounce back toward the laser, then at another point it will be bounced forward again toward the detector. Since the reflections depend on the instantaneous microscopic state of the fiber, the effective path length for these double reflections will vary continually, which means that the doubly reflected light will arrive at the detector at random times.

This means that at any moment the detector receives two signals: the direct signal plus another random (that is, uncorrelated) one from the same optical source. IIN is the result of these two signals mixing in the detector. These two signals will mix and produce beats in the RF band at multiples of the difference frequency of their two wavelengths.[8]

There will be more reflections for longer lengths of fiber, but as the fiber length increases, so does the attenuation of these extra-long-path double reflections. At shorter fiber lengths, IIN increases rapidly—approximately as the square of length. As the fiber length is increased, the rate becomes essentially proportional to length. In most cases, the maximum effect of IIN is seen at fiber lengths between 10 and 20 km.

In the case of a laser with a 10 MHz linewidth, the IIN spectrum extends up to 20 MHz (twice the linewidth). If the laser linewidth were twice as wide, the total amount of IIN would not increase, but it would be spread over twice as much bandwidth. Thus the noise impact in a given communication channel would be reduced by 3 dB. The DFB linewidth can be broadened through the mechanism of *laser chirp* by applying a constant modulating tone (sometimes called *dithering* the laser). Chirp means that the laser wavelength changes as the dithering signal amplitude increases and decreases.[9]

Since the dithering tone is independent of the applied RF, it broadens the laser wavelength even when there is only a small RF load on the laser. Note that for forward path lasers, there is always some degree of linewidth broadening

8. Since the laser linewidth is not infinitesimally narrow, even DFB lasers will generate non-zero wavelength differences in this case.

9. This is sometimes referred to as AM-to-FM conversion since amplitude modulating the laser also modulates the wavelength (frequency).

because the forward analog signals are applied constantly. On the return path, however, there is no assurance that the lasers will have RF signals applied. In the worst case there could be only one late-night subscriber sending information upstream, so the laser would be lightly loaded, and the one subscriber should expect a clear channel. Thus when linewidth broadening is helpful in reducing noise in a return path laser, dithering must be accomplished with an independent signal. This will generally be at a frequency below the return band.

For DFBs, dithering may also improve spur performance by forcing the laser to operate in several modes simultaneously.

Detector noise

The other two noise contributions occur in the fiber receiver. In the receiver's photodetector, incoming optical photons are absorbed and photoelectrons are ejected and collected as a current. *Shot noise* occurs because each photon arrives as a discrete event, so the output current from the detector will have a statistical fluctuation to it. The higher the RF signal power is at the detector, the smaller will be the effect of shot noise statistics. Thus, for a given laser, shot noise becomes more important as the fiber length increases. Shot noise is significant at all link lengths when the laser is lightly modulated (OMI $\leq$ 10%). When shot noise is the main contributor, C/N will decrease 1 dB for each additional dB of fiber loss.

Receiver thermal noise arises from the same thermal fluctuations that occur in amplifiers. As in RF amplifiers, this effect becomes important when the input signal powers are low because the thermal noise is a constant, depending only on receiver temperature. Thus, in very long fiber links, thermal noise will dominate and the C/N will decrease by 2 dB for each additional dB of fiber loss.

Figures 9.6 and 9.7 show how all of these noise contributions combine for DFB and FP lasers, respectively.

Performance over temperature

Forward lasers operate in a relatively well controlled headend environment, but return lasers must operate wherever the fiber node station is located. Thus they are usually specified to operate in ambients from –40 to +85°C.[10] The extremely high level of performance obtained in the forward path is caused, in

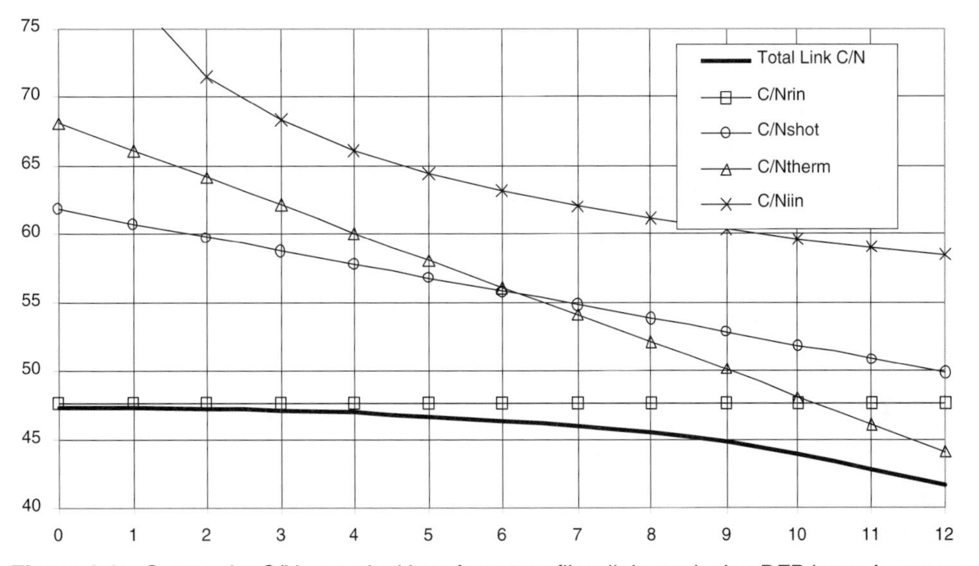

Figure 9.6 Composite C/N vs optical loss for return fiber link employing DFB laser: Laser power = 0 dBm, m = 20%, RIN = –140.

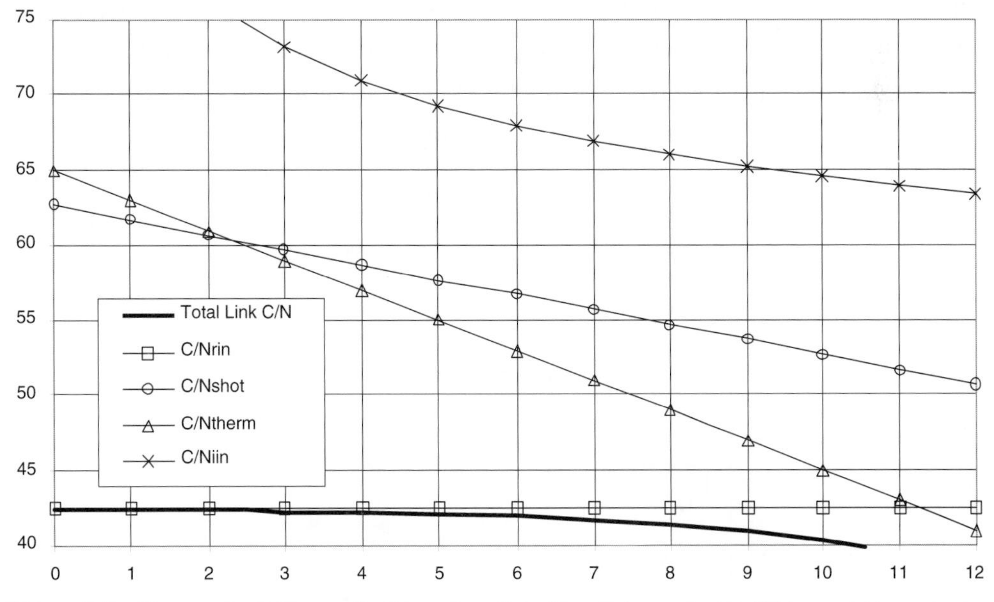

Figure 9.7 Composite C/N vs optical loss for return fiber link employing FP laser: Laser power = –4 dBm, m = 35%, RIN = –130.

10. The node ambient is usually specified to be –40 to +60°C. The laser module needs to overcome the outside ambient plus the heat generated inside the node, so its specified upper operating temperature is higher.

part, by forward lasers that incorporate *thermoelectric coolers* (TEC) to keep their operating temperatures constant. TECs are generally not used in return lasers, for two reasons. First, TECs cannot provide sufficient cooling to overcome a +85°C ambient. Second, they require considerable power, two different DC voltages (usually plus and minus 5 volts), and additional control circuitry, all of which are difficult to provide within the confines of a fiber node.

Thus the challenge has been to develop lasers that can handle the return payload without needing coolers. There are two main areas for concern as the operating temperature rises: changes in RF output level and increases in noise.

Level changes result from two effects. The laser chip itself becomes less efficient at high temperatures, which means that less light is emitted for a given current. If we think in terms of the L-I curve for the laser, this means that the slope of the optical output characteristic has decreased (Figure 9.8). We refer to this as a decrease in the laser's *slope efficiency* (milliwatts per milliamp). The closed-loop automatic power control circuit will respond by raising the bias current to bring the DC optical power back to its original value, but this means that there will be less RF power in the optical signal (unless the RF input current to the laser is given a compensating increase). This is shown in Figure 9.8, where the RF current amplitude $\Delta I'$ at 85°C is the same as ΔI, but the RF output $\Delta L'$ is considerably less than ΔL. As the slope efficiency decreases, so does the OMI.

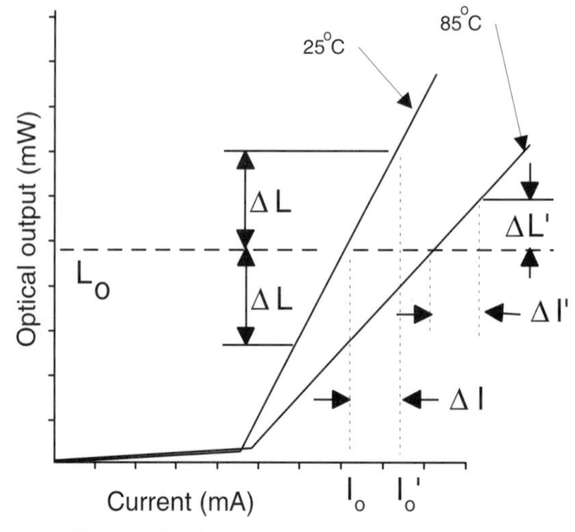

Figure 9.8 Temperature effect on laser

The second effect of temperature change is caused by *tracking error*, which relates to how well the optical power read by the rear facet monitor photodiode agrees with the actual laser output power going into the fiber through the front facet. Over temperature, the amount of error can change in an indeterminate way due, for instance, to changes in the optical alignment between laser chip and fiber end. This means that the closed loop optical power control may not actually be maintaining a constant power in the fiber.

The received RF power is proportional to both optical power and OMI. Thus it is impacted by both of these thermal effects. It is ironic that the optical transmission link appears to behave exactly opposite to the coax links. In the fiber link the electrical RF excitation doesn't change and the fiber attenuation is independent of temperature, but levels become temperature-dependent because of the active element (the laser); but in the coax link, RF inputs change and the cable attenuation varies, but the RF amplifier characteristics are largely independent of temperature.

The noise from a laser is also temperature-dependent. This is particularly true of DFBs, where it is even possible for thermal effects to cause the laser to stop operating as a singlemode device. This is because the gain curve changes with temperature, but the DFB grating does not keep pace: the grating spacing changes with temperature in a straightforward mechanical expansion, but the gain curve behaves in a much more complicated way. As a result, it is possible for the gain curve peak to move away from the wavelength that is selected by the grating. This can force the light to ignore the grating, in a sense, and cause the laser to begin operating as an FP. Since the laser design is optimized for DFB operation, it is not hard to realize that its characteristics will be marginal under FP conditions.

Last, the performance of optical isolators is temperature dependent. This generally means that their beneficial effects in reducing noise will decrease as temperatures rise. One way of compensating is to employ a series arrangement of two isolators that have offsetting thermal characteristics.

Choosing a Return Laser

It may be necessary to reassure those readers who have just read the preceding long list of problems and potential problems that it *is*, in fact, possible to obtain cost-effective lasers that fulfill their critical role in the return path. These are available from a number of equipment suppliers. The reward for your persistence will be, we hope, an improved ability (a) to tailor your laser transmitter purchases to your system needs and (b) to understand what is happening in the event that some unusual performance is observed.

We should point out, as well, that there is not really a single "best" laser for all return path applications. The latter point is made clear in Figure 9.9, which shows a DFB and an FP laser that provide comparable C/N performance.[11] This section will discuss how the various application requirements will influence the choice between competing laser solutions.

In Table 9-2 we list fiber link C/N for lasers with 0 and 3 dBm (1 and 2 mW) output powers, with three different RIN values, operated at a range of OMIs. The low RIN entries are typical of DFBs, and the high RIN of FPs.

We are now in a position to discuss lasers from the applications viewpoint.

Cost

There is not a large difference in the cost of fabricating a DFB laser chip versus an FP. The end cost of the laser transmitter is likely to remain at least a hundred dollars higher for the DFB, however, because of the need for a high-quality isolator and for added complexity in the control circuits. Due to the relatively small number of subscribers per node, this cost difference is $1–$4 per home passed, hence it is significant. The cost difference for an isolated FP will be somewhere in the middle of that range.

11. We have referenced all of these C/N numbers to full use of the return spectrum, which means that the noise bandwidth is 35 MHz, as will be made clear in the following chapter.

Table 9-2 C/N performance for various laser parameters (35 MHz bandwidth)

Row	Power	OMI	RIN	C/N at fiber loss of:			
	(dBm)	(%)		1 dB	4 dB	7 dB	10 dB
1	0	10	–150	49.4	47.4	44.2	40.0
2	0	20	–150	55.5	53.3	50.2	46.0
3	0	30	–150	59.0	56.9	53.7	49.6
4	0	40	–150	61.5	59.4	56.2	52.1
5	3	10	–150	50.5	49.1	47.0	44.0
6	3	20	–150	56.5	55.1	53.0	50.0
7	3	30	–150	60.0	58.7	56.5	53.5
8	3	40	–150	62.5	61.1	59.0	56.0
9	0	10	–140	41.3	40.9	39.9	37.9
10	0	20	–140	47.3	46.9	46.0	43.9
11	0	30	–140	50.8	50.4	49.5	47.4
12	0	40	–140	53.3	52.9	52.0	49.9
13	0	50	–140	55.3	54.9	53.9	51.9
14	3	10	–140	41.4	41.2	40.8	39.9
15	3	20	–140	47.4	47.3	46.8	45.9
16	3	30	–140	51.0	50.8	50.4	49.4
17	3	40	–140	53.5	53.3	52.8	51.9
18	3	50	–140	55.4	55.2	54.8	53.8
19	0	20	–130	37.5	37.5	37.4	37.0
20	0	30	–130	41.1	41.0	40.9	40.6
21	0	40	–130	43.6	43.5	43.4	43.1
22	0	50	–130	45.5	45.5	45.3	45.0
23	3	20	–130	37.6	37.5	37.5	37.4
24	3	30	–130	41.1	41.1	41.0	40.9
25	3	40	–130	43.6	43.6	43.5	43.4
26	3	50	–130	45.5	45.5	45.4	45.3

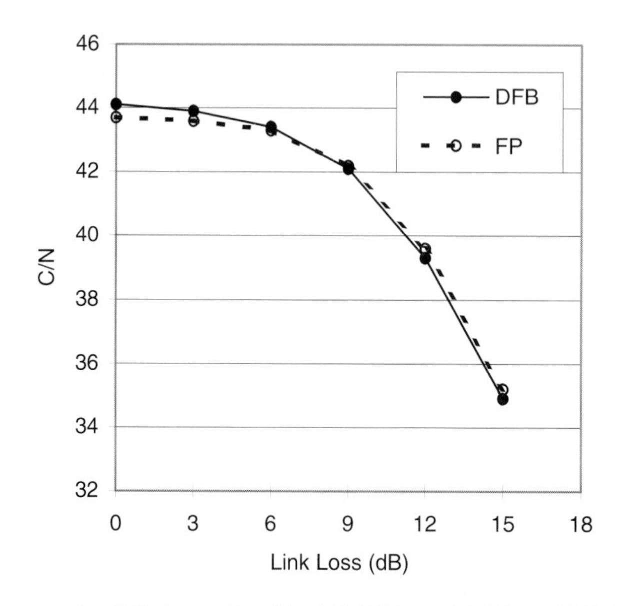

Figure 9.9 C/N curves for DFB laser (1 mW, -140 RIN, 14% OMI) and FP laser (0.4 mW, -131 RIN, 35% OMI) loaded for 35 MHz of application bandwidth

Bandwidth

As mentioned earlier, MPN in FP lasers becomes worse for high RF frequencies. This would make a DFB the clear choice for wide bandwidth returns, such as would be needed in systems utilizing block conversion (Chapter Eight).

Distance

Return fiber links are generally short since most HFC systems have been implemented—at least initially—with optical splits in the forward path.[12] Nonetheless, there are situations where longer spans are required. (We will discuss some situations of this kind in Chapter Twelve.) Scanning Table 9-2 shows that higher power helps and that the best RIN is required in such cases. For example, if a link of >10 dB needs to deliver performance better than 50 dB C/N and the OMI

12. Many HFC plants have been designed to do forward broadcasting very cost-effectively, which is appropriate, since that has been the "bread and butter" for the industry. Since forward lasers have been a major cost item, it has made sense to use higher power (10–12 mW) lasers and split the output among 2–4 nodes. This makes the actual fiber distance only 3–6 dB. Since returns are not optically combined, the return fiber loss budget will be simply that 3–6 dB.

needs to be kept below 35 percent (to provide clipping headroom), then nothing less than a 3 dBm laser with –140 RIN can succeed.

Notice that for the longer distances listed in Table 9-2 (7 and 10 dB), the low RIN FPs appear quite capable when operated at somewhat higher OMIs.

Data error rates

We mentioned earlier that spurious emissions resulting from interactions between the laser and the fiber can cause errors in higher-order modulations. Furthermore, the error rates are not as predictable as they are for the other noise sources, which are closer to additive white gaussian noise. These spurs can be reduced by incorporating optical isolators into the laser module. This works very effectively with FP lasers, but is more difficult to achieve in DFBs. In this case "more difficult" means more costly, not impossible. For DFBs, spurs at the low end of the return band are particularly persistent.

Dynamic range

The usable range of RF input powers (the *dynamic range*) for any laser is determined at the low end by the noise phenomena we have been discussing and on the high end by clipping and distortion. It is obvious that a wide dynamic range is good because it allows the laser transmitter to be operated with comfortable carrier-to-interference margins against both noise ingress (which raises the system noise floor) and signal overload (which causes intermods).

For relatively short optical links (<6 dB), DFB lasers will have larger dynamic ranges than FPs because of their lower RINs. It is possible to obtain DFBs with dynamic ranges of 15–20 dB, while FPs will be generally 5 dB lower.

The next chapter discusses how to quantify dynamic range and how to use it most effectively.

Summary

In Table 9-3, we have summarized some of the differentiators in choosing between return path laser types.

Table 9-3 Summary of performance characteristics associated with return laser technologies

Laser type	Power	Isolator	Cooling	Other
DFB	Higher powers possible	Required	Required for >2 mW	More temperature sensitive
FP	Up to 1 mW	Improves predictability of performance	Not needed	Less expensive. Limited RF Bandwidth

Combining Return Signals

When signal loading is expected to be light and communication access is only by polling (Chapter Eight), as in systems that use the return solely for set-top box returns or status monitoring, operators should be looking for ways to minimize the amount of equipment investment. One potentially appealing approach is to cut down on the number of optical receivers by optically combining return signals from several nodes before the receiver. This appealing idea should be avoided. When return signals are to be combined, the combining needs to be done in the RF (coax) domain, after the fiberoptic receiver. It is almost never a good idea to combine optical returns directly, before the receiver.

Optical combines cause difficulties in two ways: careful power balancing of the individual returns is required, and highly variable amounts of optical interference between the two signals can occur. The first effect adds complexity to plant set-up; the second effect can be devastating to performance.

The optical spectra of FP and DFB lasers (diagrammed in Figures 9.2 and 9.3) are temperature-dependent. This is because the optical "tone" of the laser cavity changes due to thermal elongation or contraction of the semiconductor material. If lasers in two different nodes are subjected to different temperatures, their optical characteristics will move relative to one another, and the amount by which the two spectra overlap will vary from moment to moment. By combining the optical outputs from the two lasers within one fiber, the two signals can mix when the wavelengths approach one another, which causes huge quantities of sum and difference beats to be generated in the return RF spectrum.

Three traces are shown in Figure 9.10 to illustrate this mixing. The figure shows the output from a spectrum analyzer that is fed by an optical receiver collecting the combined signal from two DFB lasers.[13] The lowest trace is the noise floor when the wavelengths of the two lasers do not overlap. The peaked curve in the middle is a snapshot of a moment when the two laser wavelengths coincide. The top curve is a "max hold" of the peaks, showing that there is more than a 40 dB rise in the noise floor as the beats pass through the return band.

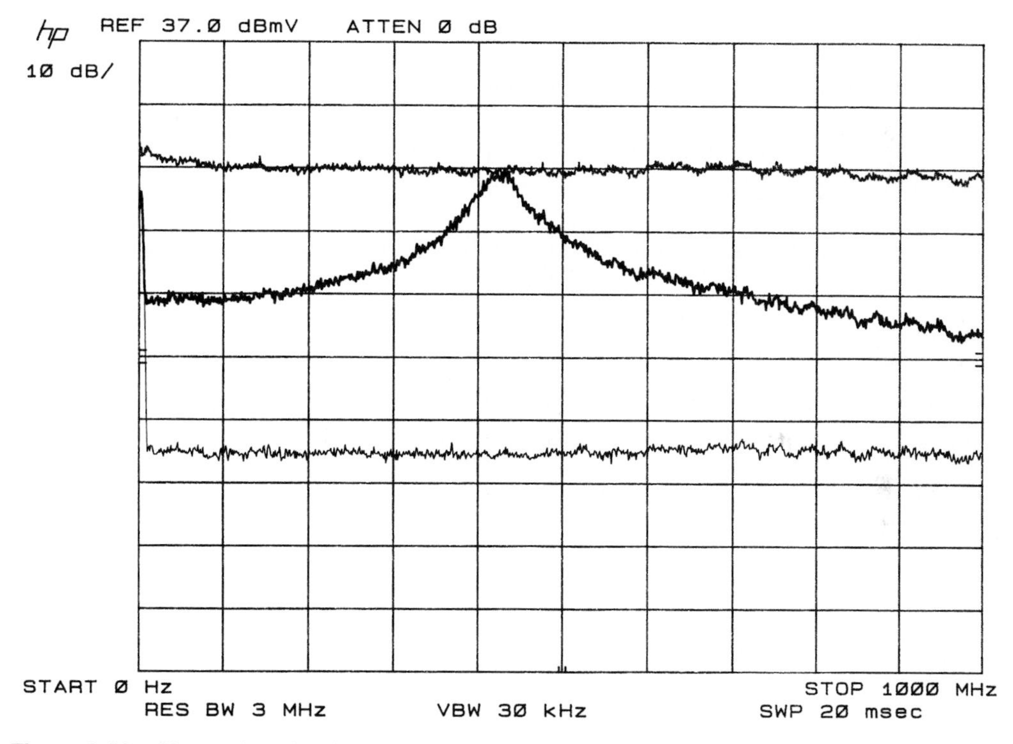

Figure 9.10 Change in noise floor as the optical spectra of two lasers overlap

The difficulties in dealing with the set-up and the noise fluctuations are likely to make the receiver equipment savings a questionable economy. At best,

13. Since the optical spectrum of an FP laser contains many wavelengths, FPs will have frequent overlaps, making optical combines even worse for FPs.

optical combining allows the operator only to delay purchasing the full complement of optical receivers needed to accommodate full use of the return band.

Summing-Up...

- Return systems are often limited by noise, clipping, and spurious emissions performance of the fiber links.
- The mechanisms for all of these phenomena can be understood.
- There is no one *best* laser for all applications, but in essentially all cases there is a suitable choice.
- Return path links should not be combined before the optical receiver.

CHAPTER 10

RF Power Allocation

A fully loaded return system will contain a number of diverse services, transmitted in a variety of modulation schemes with widely different bandwidths. As an example, a composite return signal could consist of a few 300 kHz FSK telemetry signals (status monitoring and converters), two 6 MHz bands of CDMA for PCS telephony, and 15 MHz of 16-QAM for data modems. An immediate question is: How—with this hodge-podge of signals—should we assign operating levels to each of the individual components?

In this chapter we will deal with this question and suggest an answer that is effective and easy to apply. We start by explaining that the RF power in the return path is limited by the laser. We then establish a method for determining the total power capability of the return path and conclude by discussing how to apportion that power.

Along the way we will cover two important subjects related to power allocation: (a) how power is defined and measured for digital signals and (b) how the ratio of peak to RMS voltage differs for distinct types of signals—and why that difference can be significant.

How Much RF Power Can Be Put Through the Return System?

Over the years cable engineers have been led by their experience to believe that the upper limit on RF power in a cable system is determined by distortions in the system, where by "distortions" we mean the predictable second- and third-order beats that the standard CSO and CTB tests measure. While that was true before the cable industry adopted AM fiberoptic technology, it is no longer strictly the case for the forward path[1] and it is certainly not true for the return path. The advent of lasers changed that. While distortions are still important in the forward path for determining how long an amplifier cascade can be, the limit on total RF power is set by laser clipping in both forward and return systems.[2]

Clipping demonstrated with noise notch

One way to examine the effects of clipping is to pass the signal from a broadband noise generator through a notched bandpass filter, as described in Chapter Four. The noise is a simulation of many digitally modulated carriers. The filter we have used for return band testing passes signals between 5 and 40 MHz, but has a narrow notch 50 dB deep at 22 MHz. The filtered noise, which we will refer to as the "signal," is applied to a laser transmitter at some relatively low level.[3] The carrier to noise can be measured with a spectrum analyzer[4] by subtracting the power at the bottom of the notch (which is the noise due to the fiberoptic link itself, and which we refer to as "noise") from the signal power.[5]

1. D. Raskin, D. Stoneback, J. Chrostowski and R. Menna, "Don't Get Clipped on the Information Highway," NCTA Technical Papers, pp 294–301, 1996.

2. Clipping is, of course, within the general class of what we call "distortion." Still it is worthwhile to distinguish the relatively well behaved CSO and CTB beat phenomena from the less predictable outcomes of clipping. Thus we will refer to distortion and clipping as if they are two separate items.

3. Don't be confused by the fact that our "signal" comes from a noise generator.

4. J.L. Thomas, "Cable Television Proof of Performance" (Englewood Cliffs, NJ: Prentice Hall PTR, Inc., 1995).

5. It is convenient to set the spectrum analyzer to the Noise Marker mode, which measures the noise power in a 1 Hz bandwidth. However, for depth measurements, the normal marker is sometimes more useful. This is because some noise markers average all the data in several graticles of the display, which in turn makes it difficult to make a measurement inside a notch. Regardless of the method, very heavy video averaging (or a video bandwidth $\leq$ 100 Hz) should be used so that the trace is smooth.

As the level into the laser is increased, the received signal power increases, but the noise does not. Thus the C/N increases 1 dB for each dB increase in input level, until the ratio is limited by the depth of the notch. At a certain input level, however, the C/N begins to decrease dramatically as the notch fills up with products from laser distortion and laser clipping. The notch fills at a rate that is much greater than the 2:1 or 3:1 that would be expected from second- or third-order intermodulation products.

We can also directly observe the effect of this clipping on data transmission by placing a QPSK signal in the notch at 22 MHz and repeating the test. Data from such a test are given in Chapter Four (Table 4-4). The degradation of the QPSK constellation as the signal level is increased is shown in Figure 10.1. This is a clear case where "more" is not better.

The noise notch test shows us that the performance of the return system can depend strongly on the amount of clipping in the system. The degree of clipping is a function of the *peak* power of the signal applied, which—in the case of a signal applied to a laser—is proportional to the OMI (Chapter Nine). It is unfortunate in this case that we are accustomed to dealing with average powers rather than peak powers. Power meters measure average power, and spectrum analyzers are calibrated to measure RMS levels, which correlate directly with average power.[6] The next section describes how we can determine peak powers.

Clipping demonstrated by peak voltage distribution

We have done a test in which a number of uncorrelated sinewave tones are applied to a laser. The output of the optical receiver at the other end of the fiber is sent to a very fast (35 GHz) oscilloscope. If the scope[7] is operated in the infinite persistence mode, it is possible to make a histogram of instantaneous amplitude (voltage) measurements.[8] Several million samples will accumulate in

6. The utility of the RMS concept is that an arbitrary waveform with an RMS voltage of V_{RMS} has the same power as a DC signal of the same voltage.

7. Hewlett-Packard 54120T Digitizing Oscilloscope.

8. Raskin, Stoneback, Chrostowski, and Menna, Appendix 1.

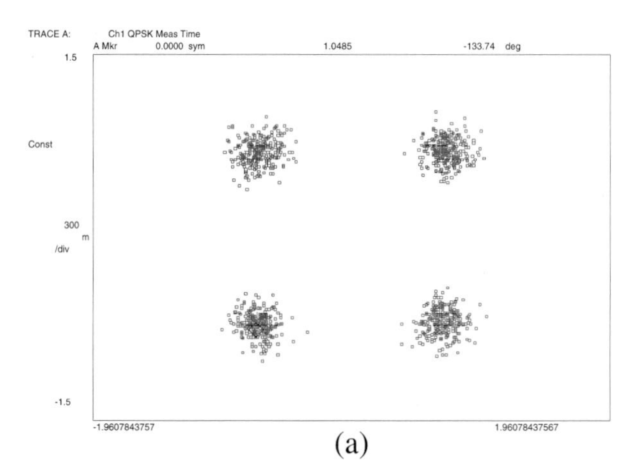

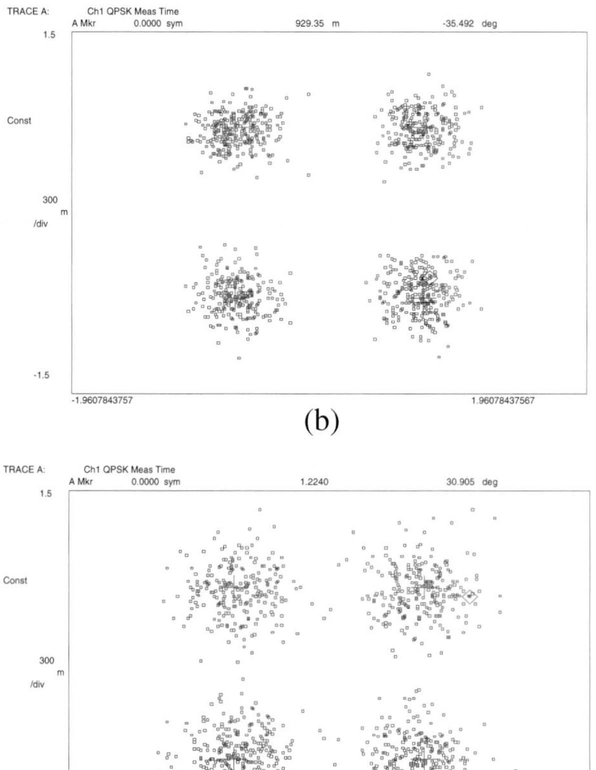

Figure 10.1 QPSK constellations at different input levels, using noise notch. (a) BER = 10-9; (b) BER = 10–5; (c) BER = 10–3

20 minutes when the scope is set to free run at a rate of 500 kHz with a timebase of 100 ps/div. Since the scope is free running at a very high acquisition rate, the distribution of dots that accumulate on the display is a representation of instantaneous voltages present in the composite signal. When the laser is operated at nominal levels, the histogram will be a normal bell curve of amplitudes equally distributed around the average. The *peak factor* (sometimes called the *crest factor*) of the input signal—the ratio of peak amplitude to RMS average—is equal to the maximum signal recorded in the histogram divided by the RMS of the signal, as calculated by the scope. The peak factor is generally expressed in dB.

Prior to the onset of clipping, the peak factor of the signal is independent of input level. As the levels are increased into the region of laser clipping, however, the histogram becomes clipped on one side. Figure 10.2 shows the accumulation of about three million amplitude points plotted above and below the central axis as a function of time (for approximately ten minutes). One can discern that the upper portion of the distribution has been clipped by the laser, which was driven several dB above normal input power. We have used this method to measure peak factors for several types of input signals, and have listed the results in Table 10-1.

Clipping is not usually caused by ingress

In Chapter Four we raised concerns about ingress of diffuse and discrete signals. One concern that we can remove is the impact of constant or slowly varying ingress on laser clipping.

Our testing has shown that it is highly unlikely for large ingress to overload the laser. Consider a laser transmitter that has a recommended total input power of 45 dBmV. Figure 10.3 shows a simulation of full 35-MHz-wide loading of the spectrum with a total payload power of 45 dBmV. Also shown is a very bad example of ingress with a total power of 31.4 dBmV. Notice that the ingress and the payload add to a total power of 45.2 dBmV. At lower frequencies the carrier-to-noise ranges from 22 to 6 dB, and at higher frequencies it is about 26 dB. There is also an interferer in the CB band which is higher than the desired signal payload. The point here is that although the ingress is so bad that a large part of the spectrum is marginally useful at best, the total loading on the laser has

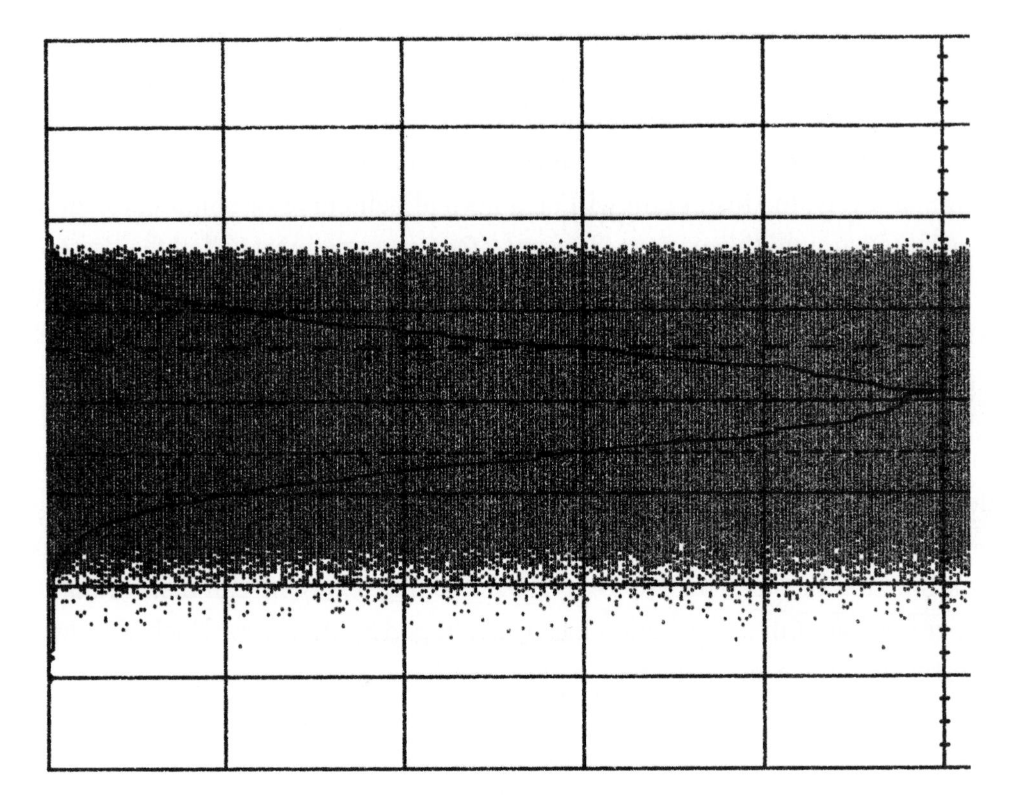

Figure 10.2 Signal amplitudes recorded with a high-speed scope over a ten-minute period for a laser that is operating above clipping

increased by only 0.2 dB. Thus, ingress affects C/N and may make certain portions of the return spectrum unusable, but it doesn't cause clipping unless it is grossly larger than the signals. This is fortunate, because when clipping is bad, it can wipe out the *entire* return band.

Impulsive ingress, consisting of noise spikes of short duration but high energy, *can* threaten the entire return band because its spectral content is so broad. In a sense, no portion of the return band remains quiet during the impulsive burst. This will result in a short duration data "outage," which may be recoverable with the use of appropriate error correction and sufficient interleaver depth (Chapter Three).

Table 10-1 Measured peak factors (peak-to-RMS ratios)

Type of signal	Measured peak factor (dB)
CW signal (sinewave)	3.2
Unfiltered Noise (5–1000 MHz)	7.8
Filtered Noise (5–40 MHz)	13.5
Modem Brand "A":	
QPSK @ 10 Mbit/sec	9.2
16-QAM @ 20 Mbit/sec	11.3
64-QAM @ 30 Mbit/sec	10.2
Modem Brand "B":	
QPSK @ 256 kbit/sec	6.7
QPSK @ 2 Mbit/sec	6.6

Notes:

The expected peak factor for a single sinewave is 3.0.

The highest peaks of the signal directly out of the noise generator are reduced somewhat due to compression in the output amplifier. After passing through a 5–40 MHz filter, the RMS has decreased because much of the signal does not pass, but the peaks are unaffected. Hence filtering results in an increase in peak factor.

What causes differences in peak factor?

Since the RMS power of a pure sinewave of amplitude E is $E/\sqrt{2}$, the peak factor of a sinewave is

$$\frac{\text{Peak}}{\text{RMS}} = \frac{E}{E/\sqrt{2}} = \sqrt{2} = 3\ \text{dB},$$

which is reasonably close to the measured value in Table 10-1. If N sinewaves of equal amplitude E, with well-defined but randomly chosen phases, are combined, the peak is equal to $N \times E$ (since that will be the value when they eventually all line up). Because the power contained in the composite is the sum of the power of the N individual sinewaves, the RMS of the composite is:

$$\frac{(V_{RMS})^2}{R} = N \times (\text{Power of one sinewave})$$

$$= \frac{N \times (\text{RMS of one sinewave})^2}{R} = N \times \frac{(E/\sqrt{2})^2}{R}.$$

V_{RMS} can then be found by multiplying the first and last expressions by R, which gives:

$$(V_{RMS})^2 = N \times (E/\sqrt{2})^2 = N \times \frac{E^2}{2}$$

$$V_{RMS} = E \times \sqrt{N/2}.$$

This means that the composite peak factor will be

$$\frac{\text{Peak}}{\text{RMS}} = \frac{N \times E}{\sqrt{N/2} \times E} = \sqrt{2N}.$$

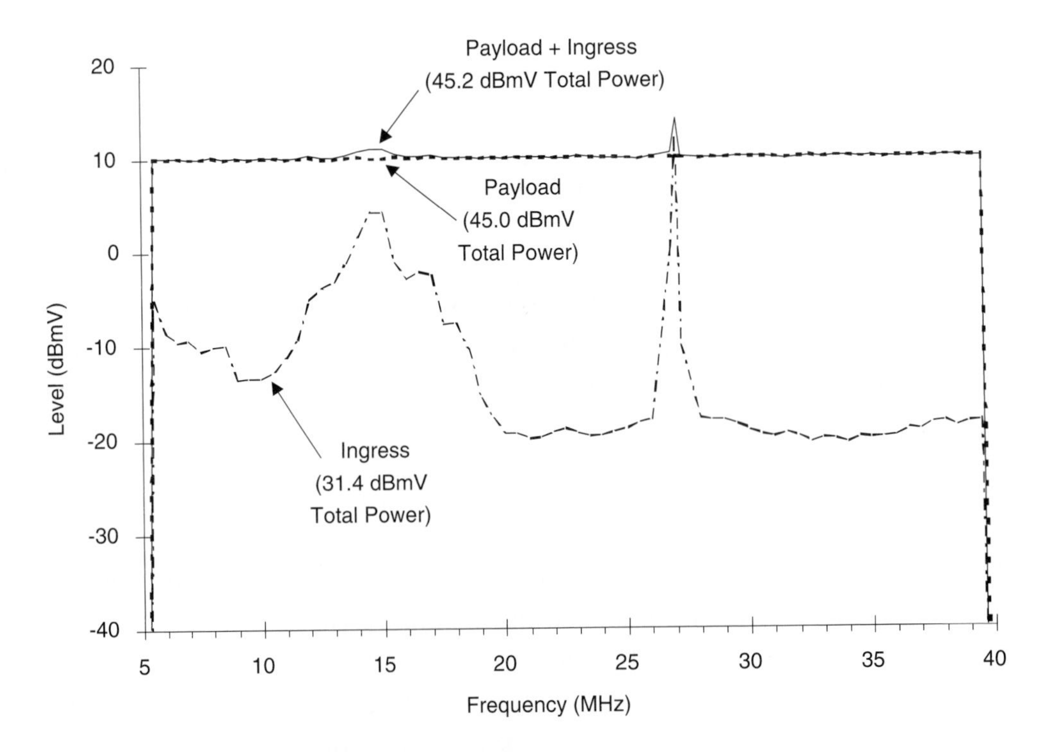

Figure 10.3 Ingress and clipping

Thus the peak factor (in dB) of N sinewaves is (3 + 10*logN). This is shown in Figure 10.4, which compares the theoretical value for phase-locked carriers with measurements of non-phase-locked signals from a multi-carrier generator. In the absence of tight phase control, the amplitudes of all of the individual waves never line up, hence the peak factor appears to level off at approximately 15 dB.

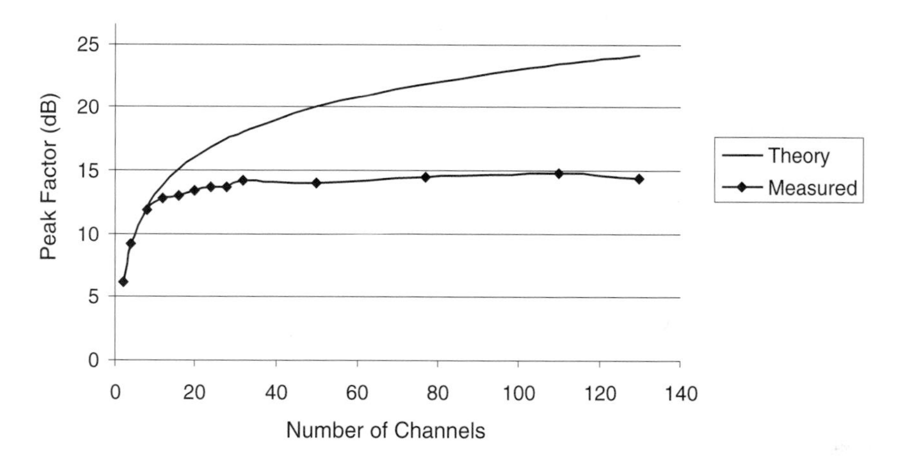

Figure 10.4 Peak factor for N uncorrelated sinewaves

Table 10-2 shows how the effective peak factor decreases with clipping (because the peaks are being clipped off). Recall from Table 10-1 that the peak factor for the return-band-filtered noise block is 13.5 dB. Within measurement error, the peak factor is unchanged after the signal passes through the fiber link at an OMI of 35 percent, but it drops as the OMI is increased. On the other hand, passing the same signal through a return hybrid amplifier shows very little clipping at an output of +65 dBmV, which is beyond the manufacturer's recommended upper operating limit. (Recall that 60 dBmV = 1 volt!) It should be noted that discrete return amplifiers have a more limited dynamic range.[9]

9. O.J. Sniezko and T. Werner, "Return Path Active Components Test Methods and Performance Comparison," 1997 Conference on Emerging Technologies, Nashville, TN (Exton, PA: SCTE) pp 263-294.

Table 10-2 Effect of clipping on peak factor

Type of signal	Measured peak factor (dB)
Laser with Filtered Noise Loading	
Equivalent of 35% OMI*	13.8
62% OMI	9.2
78% OMI	7.8
Return Hybrid with Filtered Noise Loading	
+45 dBmV total power	13.5
+65 dBmV total power	12.4
+70 dBmV total power	9.6
Return Hybrid with single CW Sinewave Loading	
+45 dBmV total power	3.2
+65 dBmV total power	3.2
+70 dBmV total power	3.2
+75 dBmV total power	3.0

* Input (average) noise power equal to the (average) power of a CW tone that gives the specified OMI

Expressing OMI in decibels

As a last item on this subject, we need to take a moment to relate OMI values, which we have been expressing in percentages, to our normal language of decibels.

Recalling that OMI is proportional to peak amplitudes, we can define a voltage E_{100} as the sinewave (CW) input that causes 100 percent OMI in a given laser. An arbitrary tone of voltage E will then cause an OMI of (E/E_{100}) percent. Thus we can define the OMI in decibels as follows:

$$\text{OMI (dB)} = 20^*\log(E/E_{100}).$$

This means that 100 percent OMI will equate to 0 dB, and 50 percent OMI will equate to –6 dB. Conversely, if we apply a signal to the laser that is 12 dB less than E_{100}, the laser OMI will be 25 percent. For reference, the first two columns of Table 10-3 list the correspondence between OMI expressed in percentages and in dB for a CW tone.

We usually characterize lasers either with CW tones or with noise blocks. This raises a problem, however, because the tool we commonly use for comparing signal strengths is a power meter or a spectrum analyzer, either of which gives us average power, rather than peak. If we were to apply a sinewave and later a noise block of the same average power, the OMIs would be very different because of the difference in the peak factor for the two signals. We have added two columns to Table 10-3 to help in such cases. They list the amount by which the average power needs to be reduced—for a pair of sinewaves and for a noise block—to get the same OMI as for a single sinewave signal.

Table 10-3 Conversion of OMI to dB

OMI (%)	OMI (dB)	Average power that gives equivalent OMI	
	Sinewave (Pk/RMS = 3 dB)	Two CW tones (Pk/RMS = 6 dB)	Noise block (Pk/RMS = 13.5 dB)
100	0	–3.0	–10.5
83	–1.6	–4.6	–12.1
75	–2.5	–5.5	–13.0
67	–3.5	–6.5	–14.0
50	–6.0	–9.0	–16.5
45	–6.9	–9.9	–17.4
40	–8.0	–11.0	–18.5
35	–9.1	–12.1	–19.6
30	–10.5	–13.5	–21.0
25	–12.0	–15.0	–22.5
20	–14.0	–17.0	–24.5
15	–16.5	–19.5	–27.0
10	–20.0	–23.0	–30.5
5	–26.0	–29.0	–36.5

Examples:

For a laser to be operated at 20% modulation takes a single sinewave 14 dB down from the 100% modulation level. The power of a noise block would have to be 10.5 dB below that[a] *in order to produce the same OMI.*

Conversely, if we applied a noise block with –14 dB power, the OMI of the laser would be 67%.

a. From Table 10-3: -24.5 - (-14.0) = -10.5

Throughout this section we have made use of the peak factors listed in Table 10-1.

Choosing the Laser Operating Point

We have taken extra time to discuss OMI because, as we will see in the next chapter, one of the first pieces of information required for plant design is the RF input level into the return laser transmitter. This, in turn, relates directly to OMI. The optimal RF input power is arrived at by balancing noise considerations, which favor high operating levels, against clipping concerns, which put an upper limit on levels.

It is clear from the preceding section that clipping and peak factor are central to determining this operating point. Our discussion up to this point can be summarized by the following statements:

1. The severity of clipping depends on the peak power applied to a laser, not on the average power.
2. The collection of signals transmitted on the return path will have a variety of peak factors.
3. As the number of different signals increases, the peak factor approaches that of noise (recall Figure 10.4).
4. Since we set levels with instruments that measure average powers, we need to design-in a headroom factor to account for signal peaking.

So how do we evaluate lasers to ensure adequate peaking headroom without introducing impractical constraints or unnecessarily complicated testing? We will describe two methods that have been suggested by cable engineers.

Noise-in-the-slot test

The first method was included in a landmark paper by Sniezko and Werner.[10] The authors apply to the laser-under-test an RF signal composed of a 35-MHz-wide band of noise with a deep notch filtered out near the midpoint of the band, similar to the noise block described at the beginning of this chapter. They measure the ratio of signal power to the power in the notch (the noise power ratio [NPR] discussed in Chapter Four) as the input to the laser is increased. At low input levels, the NPR will be equal to the C/N resulting from laser RIN, fiber, and receiver noise. As the level is increased, the NPR will improve, and at moderate levels the ratio will be limited by the depth of the filter notch. At high inputs, however, the notch will begin to fill (rapidly) with intermodulation and clipping products. The NPR can then be plotted against input level, as shown in Figure 10.5. The user can then choose a C/N that is appropriate for the applications expected—say a C/N of 40 dB—and the dynamic range (Chapter Nine) will be the difference between the high level that gives a carrier-to-intermods of 40 and the lower level that gives a carrier-to-noise of 40. The dynamic range for the laser shown in Figure 10.5 would be 19 dB for NPR = 40 and 23 dB for NPR = 35 dB, when operated with a 5 dB optical loss budget.

Sniezko and Werner point out that the dynamic range for a given laser will depend on how much of the optical loss budget is fiber versus splitting loss. Very significantly, they note that the dynamic range is essentially the same if the input load consists of QPSK or QAM signals, in place of the noise band. This indicates that the test is a relevant predictor of performance with actual data traffic. In part this is because the peak factor of band-limited noise is similar to the aggregate of a large number of data signals. It is worth mentioning that when either real or simulated data signals that include guard bands are used to load the laser, the measured dynamic range will be somewhat greater than with noise block loading. This has to do with the way the tests are set up. Setting the collection of data signals to give the same total power as the noise block will cause the individual carrier levels to be high (to compensate for the lack of signal power in the guard bands). That will make the NPR higher (by about 2 dB) for data load-

10. Ibid.

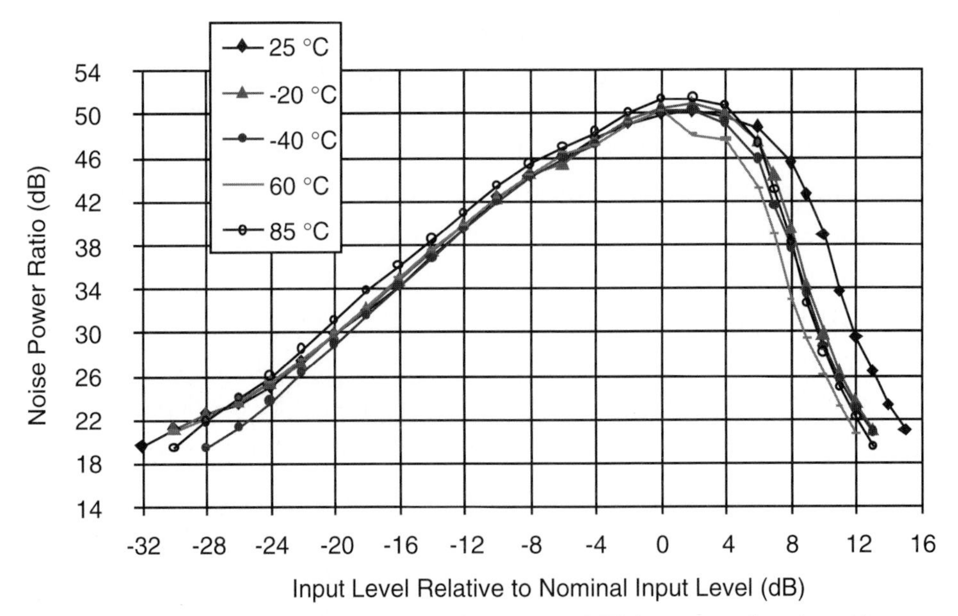

Figure 10.5 NPR performance of an isolated, uncooled DFB laser, as a function of input power and temperature (5 dB optical fiber loss)

ing than for noise loading at equivalent input powers, which means that good NPRs can be maintained as the input power is reduced. Alternatively, if the signals are set to give the same power-per-Hz as the noise, then their total power will be lower and it will be possible to raise the input further before clipping dominates. In either case the dynamic range appears to be greater with data loading. Thus we conclude that noise loading represents a worst-case test.

Their suggestion for specifying laser performance is to use this method, which they call "noise-in-the-slot," and to display dynamic range vs desired NPR level for the 35 MHz noise block, as shown in Figure 10.6. Plots for various optical loss budgets and for the full environmental temperature range should be included. Their objective is to operate lasers with at least 15 dB of dynamic range.

Two-tone test

A somewhat simpler approach, using two sinewave tones, was described to us by James Waschuk.[11] In this procedure, two sinewave tones f_1 and f_2 of equal

11. J. Waschuk (Shaw Cablesystems, Calgary, Alberta), personal communication, November 1996.

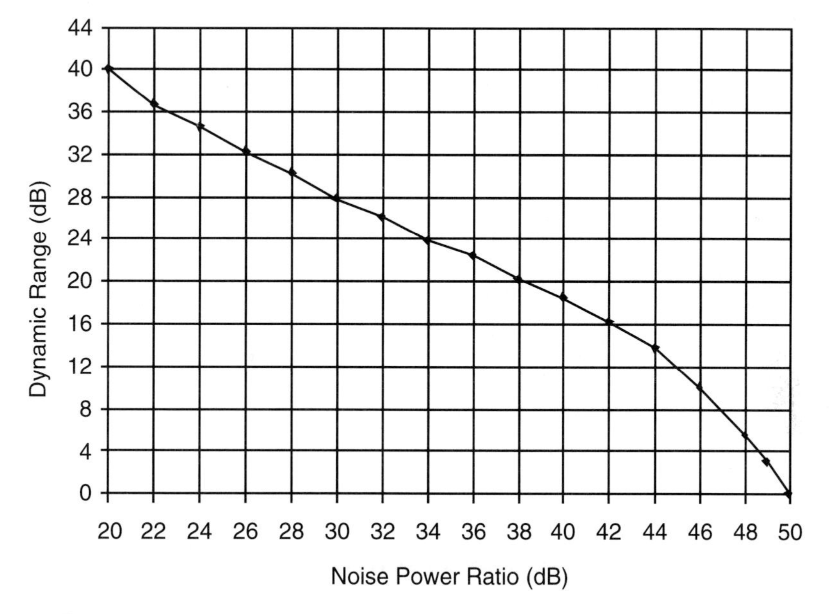

Figure 10.6 Dynamic range performance of the DFB laser shown in Figure 10.5, plotted as a function of desired NPR

power are applied to the transmitter. The power of the two tones is raised as the intermodulation products are watched on a spectrum analyzer. The $f_1 - f_2$ intermod will begin to rise faster than the expected 2:1 when the input reaches a point that we will call the *crash point* (Figure 10.7). Once the crash point power per tone E_{cr} is known, it can be used to determine the laser headroom under various loading conditions, as follows:

Since the input signal is two sinewaves, the peak factor of the test input is 6 dB.[a]
The peak voltage is equal to the RMS plus the peak factor.
In order to avoid overloading the laser, applied signals should have a peak voltage below $E_{cr} + 6$.
By Table 10-1, we can see that the highest peak factor (13.5 dB) is for a 35 MHz band of noise.
Thus an input signal power E with a peak factor no worse than filtered noise would have a peak voltage of E + 13.5, or less.

a. From the discussion earlier in this chapter, the peak factor of N sinewaves is 3 + 10*logN dB.

In order to ensure that the laser will not "crash" due to clipping, the applied signal should not have a peak voltage greater than E_{cr} + 6:

$$E + 13.5 \leq E_{cr} + 6$$

This means that a filtered noise block should not have an RMS level greater than (E_{cr} + 6 - 13.5) = (E_{cr} - 7.5).[b]

b. Alternatively, by looking at Table 10-3, we can see that a noise block needs to operate 7.5 dB lower than two CW tones in order to have similar OMI.

This analysis indicates that the laser should operate below clipping as long as the total applied power is at least 7.5 dB below E_{cr}.

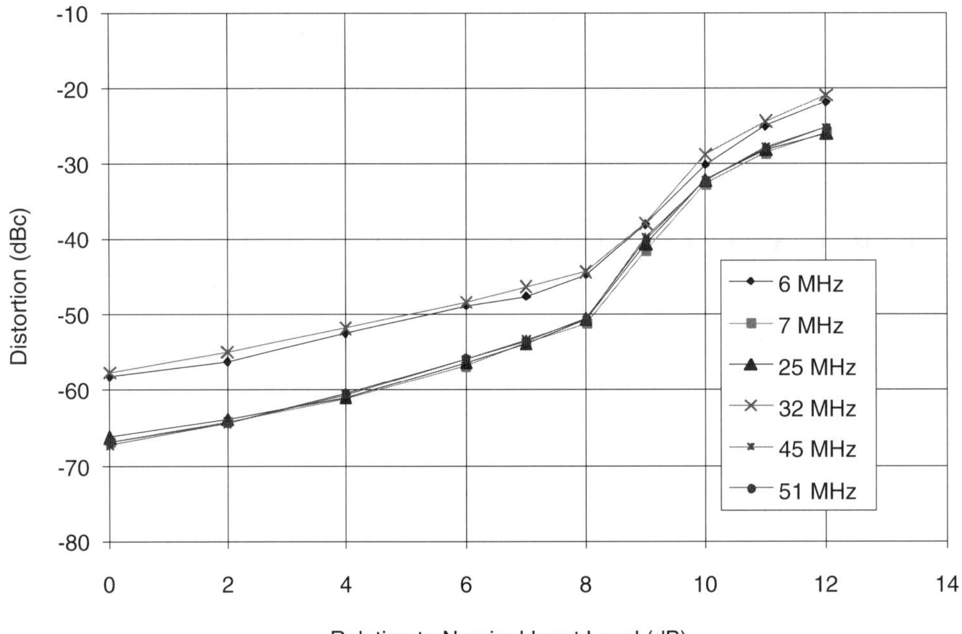

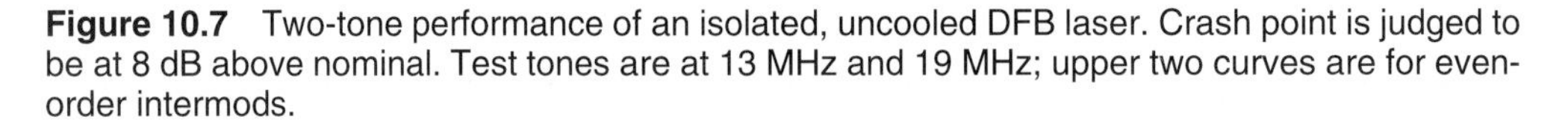
Figure 10.7 Two-tone performance of an isolated, uncooled DFB laser. Crash point is judged to be at 8 dB above nominal. Test tones are at 13 MHz and 19 MHz; upper two curves are for even-order intermods.

The setup for this method is easier than for the noise-in-the-slot method because it doesn't require special filters. It is somewhat less accurate, however, because the two tones do not represent the data loading as well as the noise block does. Specifically, a histogram of instantaneous amplitudes[12] for noise is more like data signals than is the histogram for sinewaves, as shown in Figure 10.8. We can see from Figure 10.8 that the clipping behavior for two-tone loading will be very different from multi-tone or noise loading. As the power of a two-tone load is raised and the OMI approaches 100 percent, the laser will be driven into clipping abruptly and intensely. The onset of clipping can be expected to be much more gradual for multi-tone and noise loads, due to the gaussian tails in their histograms.

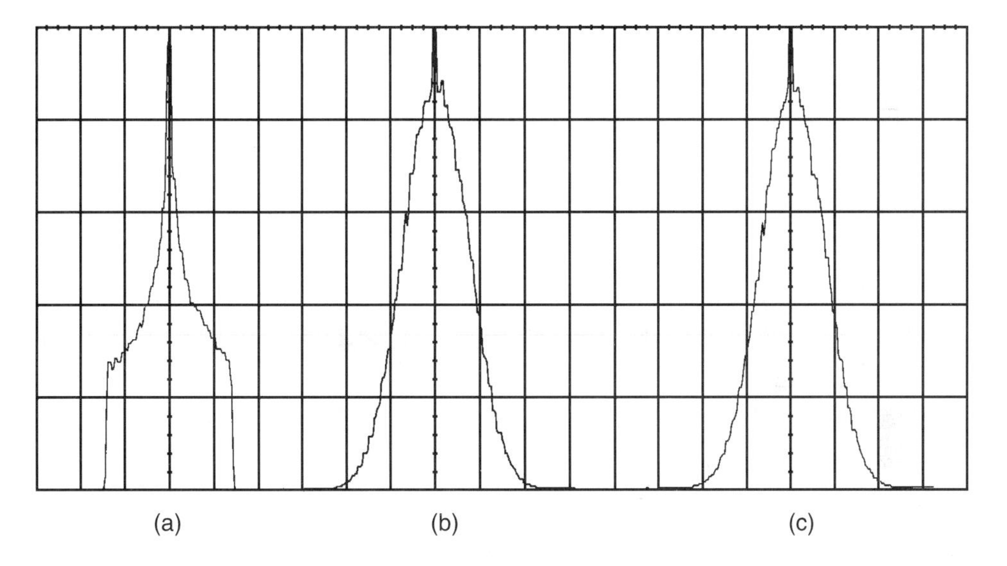

Figure 10.8 Comparison of amplitude histograms for (a) two sinewaves, (b) seventeen QPSK signals, and (c) band-limited noise, all at the same total power. Note that the two sinewaves show a well-defined peak voltage (the right and left extremes of the distribution), but the seventeen QPSK signals and the noise approach a gaussian distribution.

We have recommended that when two tones are used for setting up laser transmitters normally rated at 45 dBmV total power, the power of the tones should be 45 dBmV each. While this makes the power 3 dB higher than rated, it makes the peak a little more representative of actual full-band loading. The peak

12. This is sometimes referred to as the probability density function (PDF).

of the two tones is 54 dBmV,[13] while the peak for 35 MHz of noise at 45 dBmV average power would be 58 dBmV.

How Is RF Power Allocated Between Applications?

Now that we know how much total power can be applied to a laser transmitter, how do we decide how to share that power between applications? Several approaches are possible for properly assigning all the data channels to the total available power, but for reasons we hope to make clear later in this section, assigning power on a constant-power-per-Hz basis is suggested.

The easiest way to implement this is first to divide the total available power into 1 Hz increments. The choice of 1 Hz is convenient because many spectrum analyzers have a feature that will show noise in a 1 Hz bandwidth. Then the allotted power per Hz is assigned to each channel, based on the bandwidth occupied by the channel. For instance, if channels are spaced 1 MHz apart, they would each receive 10*log(1 MHz/1 Hz) = 60 dB more power than the per Hz value. Consider the following example for a 5 to 40 MHz return:

Total power to transmitter:	45 dBmV
Total payload bandwidth:	35 MHz ≈ 75 dB[14]
Power per Hz:	–30 dBmV/Hz[15]

In Table 10-4, we allocate this power to three multi-channel applications.

Note that the power allocated to a channel is based on the channel's spacing rather than its noise bandwidth. This allocates to the channel the power that would otherwise be reserved for guard bands, which carry no payload. For instance, a particular service might have a 128 kHz noise bandwidth, but deploy channels centered 200 kHz apart. If the power allocation for the channel were to be based on 128 kHz, some of the power potentially available at the laser would not get utilized. From Table 10-4 we can see that this procedure uses all of the laser power. Multi-tone communications schemes such as OFDM (Chapter

13. 48 dBmV total power plus 6 dB peak factor for two sine waves.

14. $10*\log(35*10^6) = 75.4$

15. Power per Hz = 45 dBmV – $10*\log(35*10^6)$ = -30.4 dBmV/Hz

Table 10-4 Examples of power per Hz allocation

Type of service	Channel spacing	Power per channel (dBmV)	Number of channels	Total bandwidth (MHz)	Total power (dBmV)
Modems	192 kHz	23	36	7	38
Telephony	2 MHz	33	10	20	43
Other	1 MHz	30	8	8	39
TOTAL				35	45

Three) should be calculated in the same manner, with power allocation based on the total bandwidth used, rather than that of an individual tone.[16]

The plot in Figure 10.9 shows how this fully loaded return path would look on a spectrum analyzer. Note that a noise marker would indicate –30 dBmV/Hz. Also note that the displayed level of +10 dBmV correlates to a resolution bandwidth setting of 10 kHz since

$$-30 \text{ dBmV/Hz} + 10{*}\log(10 \text{ kHz}) = +10 \text{ dBmV}.$$

A zoomed-in view of the 14 to 20 MHz band (Figure 10.10) shows the noise floor of the system. These figures show what a spectrum analyzer would show for a C/N of 42 dB. Note that the 42 dB delta can be seen directly on the display. This is an advantage of using the constant-power-per-Hz method.

Channel power allocation on a constant-power-per-Hz basis has at least five positive aspects:

- All services have the same carrier to noise, regardless of their bandwidth.[17] There is no need to specify a particular bandwidth when measuring C/N.
- When properly set-up the spectrum will look flat when viewed on a spectrum analyzer, as in Figure 10.9.[18]

16. If the example of Table 10-4 included an OFDM service with a full bandwidth of 2 MHz, the power allocation would be -30 dBmV/Hz + 10*log(2*10^6) = -30 + 63 = +33 dBmV. If that signal consists of 100 tones, the power per tone would be +13 dBmV.

17. Assuming equal amounts of ingress in all channels.

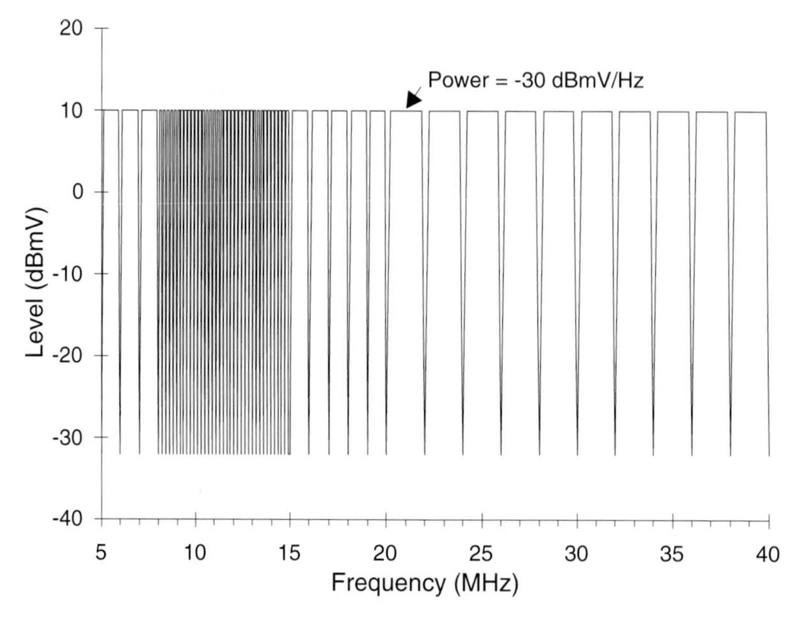

Figure 10.9 Spectrum analyzer plot

- Power allocation is simplified. The power assigned to any service is easily calculated, based only on the bandwidth of that service.
- Very narrow channels that will be subjected to less ingress energy are allocated less total power. Broad channels that are subject to more ingress energy get to use more total power.
- Return path power capacity is automatically allocated for services that will be added at a future date.

One needs to weigh these attributes against the disadvantages:

- Providing the same C/N for every service may not be the most effective power allocation if the various services have widely different C/N requirements.

18. Since most digital modulation schemes have a noise-like spectrum, the actual level on a spectrum analyzer display will depend on the resolution bandwidth setting of the analyzer. If all channels are operating at the same power per Hz, however, they will all appear at the same level on the display. (See Appendix A.)

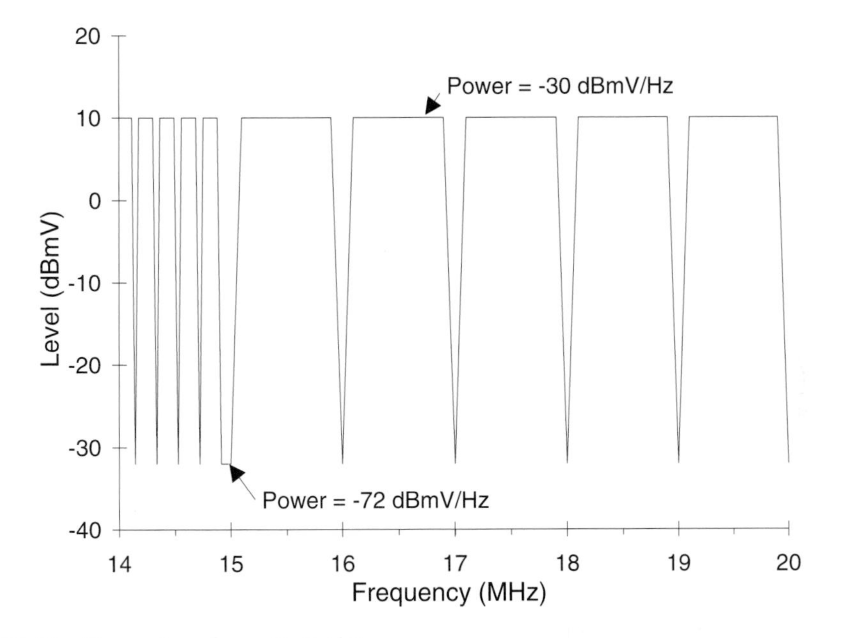

Figure 10.10 Expansion of spectrum plot

- Some of the in-home application transmitters may not have high enough output power capability to achieve the required level in the return plant.
- It is tempting not to hold power capacity in reserve for future applications, which may never be implemented.

Our suggestion is to adopt the constant-power-per-Hz methodology as a starting point because it simplifies a number of set-up issues, as will become more apparent in the next chapter. On the other hand, it should not be applied dogmatically, since there will be particular cases where it is not appropriate, as we will also illustrate in the next chapter. An example would be the use of the return path to carry a video channel along with a small amount of data. In that case, a C/N of at least 50 dB is required for the video, which can be achieved only by overallocating power to the 6 MHz video channel, at the expense of the data. In practice, of course, this will mean that there will be very little power remaining for data. In turn, the constant-power-per-Hz method—applied to the remaining bandwidth—is likely still to be relevant.

Summing-Up...

- The limit on return path RF power is determined by laser clipping.
- All lasers have similar clipping characteristics.
- Simple methods exist for determining a useful operating point for a given laser.
- A good basis for allocating power between applications is constant-power-per-hertz, although other methods can be used.

CHAPTER 11

Return Plant Design and Setup

In the preceding chapters we have discussed a wide range of technologies relating to the return path and to the applications that flow on it. It is now time to put all of this information to work. In this chapter we will discuss how to choose values for the key return path design parameters and then how to align the plant to those numbers. This will include the following topics:

- What operational questions do I need to answer before I get started?
- How do I determine the optimal powers[1] for operating the RF plant and the laser transmitter?
- What is the preferred reference point for the RF plant?
- What is a practical procedure for setting up the return plant?

The chapter deals first with the design methodology and then with the implementation of the design during plant setup.

1. In this chapter our focus will switch frequently between the individual return application signals and the combined signal for all the applications. We distinguish between these by using the word "power" to refer to the total power of the composite signal and "level" to refer to the power allocated to an individual application signal.

How Return Path Design Compares with Forward

For those readers who are accustomed to forward path design and setup, some of the procedures we are about to describe may seem unusual or even cumbersome. This is not actually the case, however. To clarify this point, we start by discussing forward path alignment. In that way we will be able to point out the parallels between the process for the forward and return.

In all cases the objective of the design is to determine the most efficient operating points for the equipment throughout the plant. One feature of HFC networks is that the fiberoptic and coax portions can be considered somewhat independent of one another until they are interconnected at the fiber node (Figure 11.1). This allows each subsystem to operate at its optimal performance point, which improves overall system efficiency.

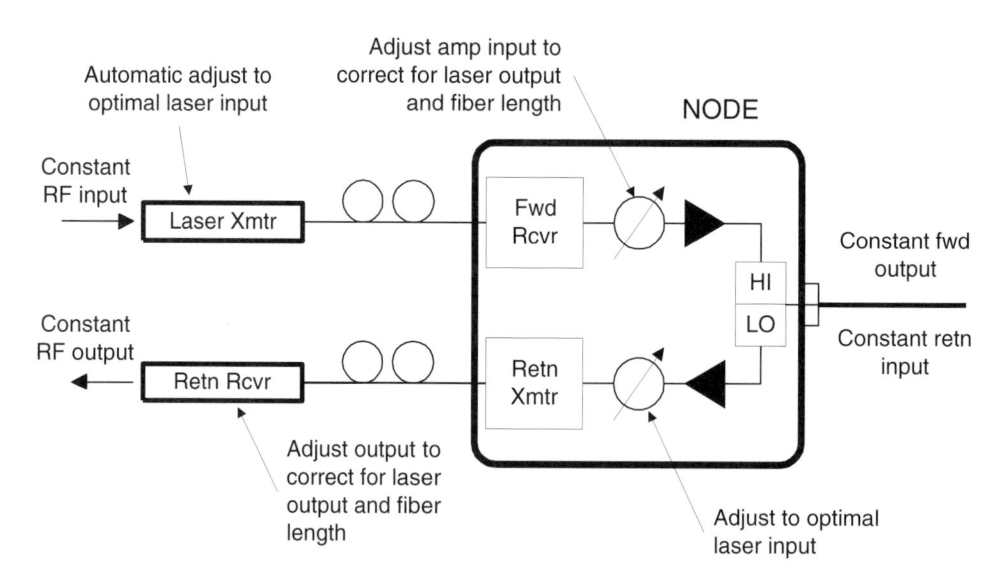

Figure 11.1 Simplified block diagram of fiberoptic system and node

Let's sketch out what we are trying to accomplish in each step of the plant alignment, using Figure 11.1.

Forward plant

A constant RF input power feeds the laser transmitters in the headend. Inside the transmitter unit, that power is adjusted by fixed attenuators and *automatic gain control* (AGC) to be at the point that optimizes laser noise and clipping performance. At the node, the output of the forward receiver will depend on the laser power and the length of the fiber link. Thus the receiver output must be adjusted—by a combination of padding and AGC in the node—to the level determined to be the proper input for the RF section of the node. The RF section is then adjusted to the output levels that give optimal noise and distortion performance for the RF plant. Usually the outputs of all subsequent amplifiers will be set to the same levels. Thus we say that the plant is aligned to *unity gain*, meaning that the gain factor between a common reference point—the *unity gain point*—at each amplifier equals one. As just indicated, for the forward plant the unity gain point is the amplifier output.[2]

Return plant

Appendix B1 explains the importance of unity gain and our reasons for recommending that the unity gain point for the return direction be the same physical port as for the forward, except that now we're thinking of it as the return input rather than as the forward output. The plant will be setup so that the RF power at the return input port will be the same for all amplifiers and nodes.[3] In the next section, we will describe how to choose this power, based on the power capability of in-home application transmitters and on plant losses. As discussed in Chapter Nine, the dynamic range of the return laser transmitter will be optimal at a particular RF input power. Thus, the levels within the node need to be

2. Some amplifiers have an internal plug-in splitter or directional coupler between the diplex filter and the output ports. This permits the signal to be distributed along two paths, avoiding the weatherproofing problems associated with an external line splitter. For reasons that are explained in Appendix B1, this convenient feature causes a bit of a problem for design. These problems are avoided—and the unity gain principle is maintained—if we think of the internal splitter or coupler as if it were external to the amplifier unit. The same mental correction is necessary for the return direction.

3. The fact that the forward signal is always present, whereas the return signal is not (since it consists largely of data bursts coming from homes) means that the sweep technician will need to inject signals when setting-up the return path, as we will discuss in detail later in this chapter.

adjusted to that value. Last, the output of the headend fiberoptic receiver needs to be adjusted appropriately for differences in fiber loss and laser output.

As we will see, at each station both the forward and the return setup procedures aim to compensate for the characteristics of the segment of plant between that station and the adjacent one closer to the headend. The required adjustments are summarized in Figure 11.1. By considering the figure, one should be able to appreciate the parallelism between the forward and return design steps.

Choosing Design Levels

Our task is to choose the optimal operating points for the fiberoptic and coax portions of the return path. More specifically, this means determining the input powers to the laser transmitter and to the amplifier and node station ports. To make these decisions we will use some of the topics covered earlier:

- Determining the proper operating point for a return laser (Chapter Ten)
- Allocating RF power between applications (Chapter Ten)
- Adding up plant losses and loss variances (Chapter Five)

Laser transmitter power

Chapter Ten described how the transmitter vendor or the operator arrives at a recommended operating point. For example, the relevant information for some of our company's products is shown in Table 11-1. We will assume that the reader has access to similar data for other vendor's products.

Plant power

Once the laser model is chosen, our only remaining design task is to decide the RF plant power. The aim is to make it as high as possible in order for the desired signals to be far above ingress.[4] Since the return path signals originate in subscribers' homes, we need to find out what power can be delivered to an

4. In Chapter Ten we noted that hybrid return amplifiers are unlikely to be driven into clipping. This makes the level-choosing goal in the RF plant exactly opposite from what it is in the fiberoptic link. For the laser, we are trying to ensure that the input does not go too high; for the RF plant, we need to assure that the levels are as high as they can be!

Table 11-1 Recommended input powers for NextLevel return transmitter products

Model name	Description	Input power for 35 MHz data loading (dBmV)
AM-RPTD	FP laser for SX node	20
AM-RPTV1	Isolated FP for SX node	25
AM-RPTV4	Isolated, cooled DFB for SX node	25
AM-MB-RPTD	FP for MB & BTN nodes	40
AM-BTN-RPTV1	Isolated FP for MB & BTN nodes	45
AM-TC-RPT	Thermally compensated, isolated FP for MB & BTN nodes	45
SG2-DFBT	Isolated DFB for Stargate nodes	15
SG2-IFPT	Isolated FP for Stargate nodes	15

Note: The wide range of input powers shown in this table is due primarily to differences in gain within the transmitter modules and node stations and only slightly to distinctions between the lasers themselves.

amplifier station's return path input when the in-home application equipment is transmitting at full power and the signal is traveling through the maximum expected plant loss.[5] To do this, two types of information are required:

- The maximum plant loss between the in-home transmitter and the amplifier
- The maximum specified transmitter output level and occupied bandwidth (for each application).

Evaluating maximum plant loss

In Chapter Five, we described the different causes of signal attenuation in the plant between an in-home transmitter and an amplifier return input port. Summary Table 11-2 can be used as a template or check-list to compile actual values

5. We are assuming that all of the transmitters are remotely controlled to turn up their power whenever the level received in the headend drops below the prescribed lower limit. This is referred to as *long-loop AGC*. We need to ensure that the plant is set up so that the long-loop AGC causes the transmitters to operate as high as possible, without running out of head-room.

relevant to your plant. We have inserted approximate numbers for the purpose of illustration. Remembering that our goal is to find the maximum attenuation hurdle that a transmitter needs to overcome, we should concentrate on the "worst case" column. Note that we have listed not only the losses from home to nearest amplifier but also the increases that might occur between the farthest amplifier and the headend receiver. These "system variances" account for

- dips in the return spectral response,
- increases in cable loss due to temperature rises,
- inaccuracy in compensating at the headend for different fiber lengths, and
- imperfection in achieving unity gain between amplifiers.

Table 11-2 Plant losses

Description	Typical Loss (dB)	Worst Case (dB)	Comments
Single span			
In-house cabling & splitting	4	8	Amplifier required if there are more than 2 splits in home
Drop cable	2	3	@ 40 MHz
Tap & feeder cable	20	27	Home closest to amplifier has highest loss
Span subtotal	26	38	
System variances			
Flatness	1	2	No flatness adjustments in amplifiers
Thermal change	3	7	Includes thermally compensated laser
Optical link length	1	2	Assumes length compensation during setup
Alignment accuracy	0	2	Inaccuracy of meters
System subtotal	5	13	
Total	**31**	**51**	

Application transmitter power and bandwidth

Now that the loss is known, we need to determine the guaranteed maximum output level of the transmitter. In a sense, we are looking for the "weakest" transmitter since that one puts a limit on the power for all applications.

Many set-top boxes can transmit at as much as +60 dBmV. However, just knowing the output level of the transmitter is not sufficient information. Remember that the return path consists of a variety of services that may have different bandwidths, and we need to allocate the power between them. In Chapter Ten we suggested that this allocation be done on a constant-power-per-Hz basis. We will apply that method in this chapter. This means that our definition of "weakest" is "smallest upper power limit per Hz of occupied bandwidth."

One fine point needs to be raised. If an in-home transmitter has a maximum power capability of +60 dBmV, but the headend controller has a power measurement accuracy of only ±2 dB, then there can be no assurance that the transmitter will ever be instructed to go all the way to its upper limit. In that case, one has to consider +58 dBmV as the actual top power for this transmitter. It may also be useful to determine the maximum assured output of the transmitter, based on the tolerance listed in its output power specification, since this will be less than the nominal value for maximum output power.

As in Chapter Ten, we calculate power-per-Hz logarithmically:

Power per Hz = (Power in dBmV) - 10*log(Bandwidth in Hz).

Let's go through the process with a sample system with three return path applications.

Service 1 (Modem)
The cable modem transmitter that we are deploying has an upper power limit of +53 dBmV and an occupied bandwidth[a] of 1 MHz. The power per Hz for that transmitter is +53 – 60 = –7 dBmV/Hz. If there is 45 dB of signal attenuation in the plant, then the amplifier return input port receives –7 –45 = –52 dBmV/Hz from that transmitter.

Service 2 (Set-top converter)
Our set-top transmitters have an upper power limit of 60 dBmV and a 250 kHz occupied bandwidth. Following the same calculation gives a power per Hz of +60 –54 = +6 dBmV/Hz. Since the plant attenuation is the same, this transmitter would be able to hit the amplifier at –39 dBmV/Hz.

a. Remember to use the occupied bandwidth or channel spacing, not the receiver's noise bandwidth

Service 3 (PCS)
We are also deploying PCS telephony with a +50 dBmV upper transmit level and a 6 MHz occupied bandwidth. This gives a power density of +50 – 68 = –18 dBmV/Hz. This service is connected directly into the coax hardline with a –12 dB directional coupler and has only 3 dB of additional plant loss, so the transmitter can deliver –18 – 15 = –33 dBmV/Hz to the station port.

Choosing the plant levels

We summarize the applications for our sample system in Table 11-3. This will help us decide on the optimal RF plant power.

Table 11-3 Plant level calculation (5–40 MHz)

	Cable modem	Set-top converter	PCS
Guaranteed upper transmit level (dBmV)	+53	60	50
Occupied bandwidth (MHz)	1	0.250	6
Transmit power density (dBmV/Hz)	–7	+6	–18
Maximum plant attenuation (dB)	45	45	15
Power density at port (dBmV/Hz)	–52	–39	–33
Power (dBmV) at port if all signals were at this application's power density	+23	+ 36	+42

Last row is calculated by adding $10*\log(35 \times 10^6)$ to entries in row above.

It is clear that the cable modem in our example is the weakest transmitter since the maximum power density that it can deliver to the port is lower than that of the other applications. By our method we would then say that all applications should be run at a power density of –52 dBmV/Hz. If that power were utilized across the full 35 MHz,[6] the total power at the port would be $-52 + 10*\log(35 \times 10^6) = 23$ dBmV.

If we look back at Table 11-3, however, this may seem a little extreme, since aligning to a power density of –52 dBmV/Hz means that the set-top converters would be instructed to transmit at levels 13 dB lower than they otherwise

6. For example, by two return bands of PCS, twelve bands of cable modems, one band for converters, and 10 MHz held in reserve for future applications.

could. That would be unfortunate, since it might mean that the set-top signals would not be high enough to overcome ingress.[7] The alternative—basing the design on the set-tops and operating with +36 dBmV at the port—would likely be disastrous for the modem system, however. There would be many homes where the modem transmitters could be operating at their full power capability, but still not generating enough signal to satisfy the requirement at their application receiver.

It is now up to the system designer to decide how to deal with the potential C/N reduction. For set-top converters using a polled data protocol, for instance, it may be a major concern since almost unlimited transmission re-tries are possible. On the other hand, it is very common for set-top transmitters to operate in the bottom portion of the return frequency band, where we expect ingress to be the worst.

Let's explore two possible approaches to improving the situation.

The first is to realize that the set-top may be the only application operating between 5 and 10 MHz. If there is no intention to add further services in that portion of the return spectrum, then we can allocate to the set-top all of the power that would be used by 5 MHz worth of applications, rather than just 250 kHz. That would mean that the set-tops could operate at 13 dB[8] higher levels without adding to the total power expected for the RF plant. The total input power at the amplifier is still +23 dBmV.

The second is to add external amplifiers to the relatively weak cable modem transmitters so they will not drag down all of the other applications. This is not as bad as it might seem because not all of the modems will actually need amplifiers. Only those working against the worst-case plant loss need the added gain.[9] A survey of the plant for example, might show that a relatively small per-

7. The PCS signals would have to be lowered even more (by 19 dB). Since implementing the low-band blocking filters discussed in Chapter Eight would allow these signals to operate in a somewhat "cleaner" environment, this might not be a problem.

8. 13 dB = 10*log(5 MHz / 250 kHz) = 10*log(20).

9. Another way to achieve the same effect is to run a separate drop directly to the unit with the weak transmitter, thus reducing the maximum plant attenuation for that application.

centage of homes are subject to highest tap and cabling losses. A 10 dB amplifier would allow the port input power to be set for +33 dBmV. Note that the amplifier needs to be located immediately in-line with the modem. If it were put at the side of the house so that all of the in-home applications were amplified, it would not have the desired effect.

We can summarize the effects of these changes by recasting the allocation table (Table 11-4). For each of the three scenarios, we have started at the top, just as before, by calculating the power density at the port for the weakest transmitter (the cable modem). We then proceed from that power density in a backward direction to calculate the transmitter power for each of the other two services. In the second scenario we have given the set-top converters the full 5 MHz bandwidth. In the third scenario we have added external amplifiers where needed for the cable modems and accounted for them by reducing the plant attenuation by 10 dB (only for the cable modem transmitter). We have highlighted the numbers in the table that change in the second and third scenarios. Both of the new designs appear to be workable.

Remember that one feature of the constant-power-per-Hz method is that it ensures that adequate power is held in reserve for future applications. We have calculated power densities as being uniform for the full 35 MHz even though our three present applications may not fill the whole return band. By assuming a 35 MHz band, we have assured that power is available for new uses that will be added later. On the other hand, if any portion of the band is certain not to be used, then a smaller bandwidth should be applied in calculating power density. For instance, an operator may decide that it is not cost-effective to tighten up the plant to avoid low-frequency ingress below 15 MHz. In that case, additional power could be allocated to applications that have underutilized transmitter power, such as PCS in our example.

Finally, recall that while ingress may cause certain communication channels to have unusably low C/N, it is unlikely to affect total power, as noted in Chapter Ten. Therefore it is unnecessary to allocate a portion of the return band power budget to ingress.

Table 11-4 Alternate plant level scenarios

Scenario	Original			5 MHz for STB			Amps for CM		
Application	CM	STB	PCS	CM	STB	PCS	CM	STB	PCS
Maximum transmit level (dBmV)	+53			+53			+53		
Occupied bandwidth (MHz)	1			1			1		
Transmit power density (dBmV/Hz)	–7			–7			–7		
Maximum plant attenuation (dB)	45			45			**35**		
Power density at port (dBmV/Hz)	–52	–52	–52	–52	–52	–52	**–42**	**–42**	**–42**
Maximum plant attenuation (dB)		45	15		45	15		45	15
Power density at transmitter (dBmV/Hz)		–7	–37		–7	–37		**+3**	**–27**
Allocated bandwidth (MHz)		0.25	6		**5**	6		0.25	6
New upper transmit level (dBmV)		+47	+31		**+60**	+31		**+57**	**+41**
Total power at port (dBmV)		+23			+23			**+33**	

Last row is calculated by adding 10*log(35 x 10^6) to power density at port.

Putting the Pieces Together

By determining the input powers for the laser transmitter and for the amplifiers, we've now achieved the two major steps of the return path design. We now have to make sure that these two determinations are consistent with the internal design of the node—where the two meet. Let's choose two particular values, 33 dBmV for the port input power and 45 dBmV for the laser input power, to illustrate how this is done. We'll use the dual-port node diagrammed in Figure 11.2.

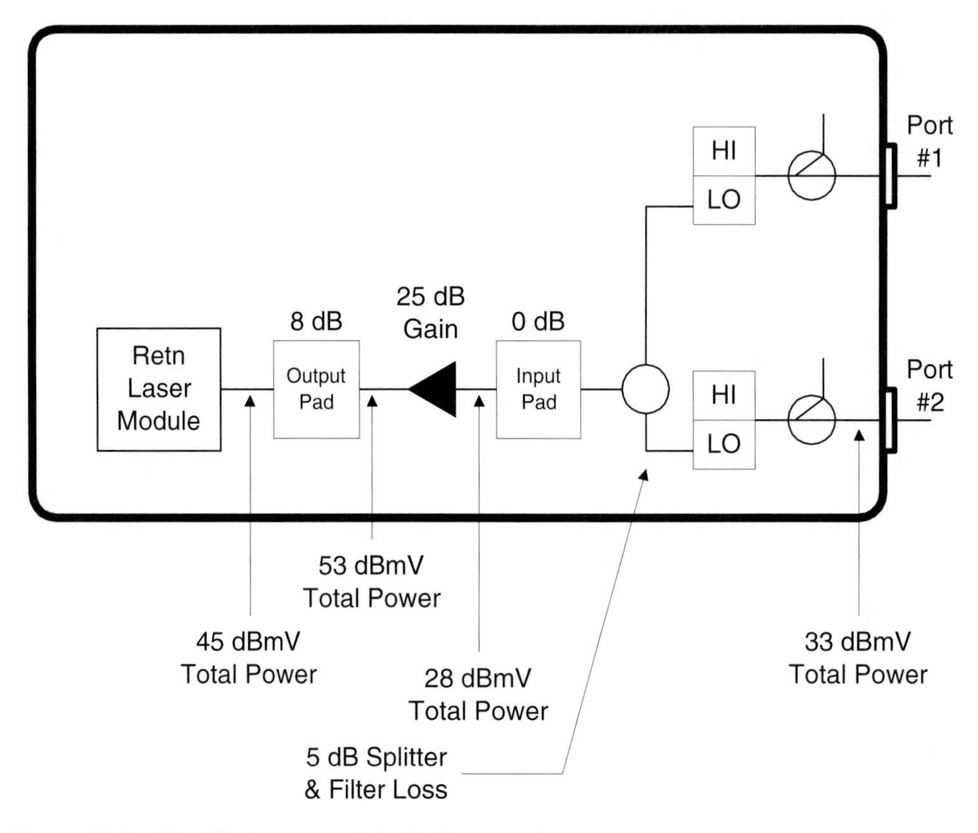

Figure 11.2 Simplified return path in dual-port node

Two positions are provided for plugging-in *pads* (fixed-value attenuators) to make the gain appropriate. In this example a total of 8 dB is required, as shown in the figure, to match the 33 dBmV port input power to the 45 dBmV laser input.[10] The maximum gain available between the port and the laser is called the *node return full gain*. In our example, the node return full gain is 25 – 5 = 20 dB.

$$\text{Pad value} = \text{Port input} + \text{Node return full gain} - \text{Laser input}$$

The *node return gain* is the gain from port to transmitter input after padding, 12 dB in this example.

10. (33 dBmV port input) – (5 dB path loss) + (25 dB amplifier gain) – (8 dB pad) = 45 dBmV.

Higher node return gains may be used in cases where the power delivered to the port is lower than desired, due either to limited application power levels or less than full usage of the return band. Recall that the node is to be padded so that the proper total power is applied to the laser transmitter. Therefore, if the applications can deliver only 30 dBmV to the port, then the pad value would be 5 dB, in the above node example. This has the effect of maintaining the C/N of the fiberoptic link for each of the applications.

Pad and equalizer locations

For people used to forward path setup, it might appear strange that we have put all of the attenuation after the amplifier hybrid. This is another reminder that amplifier distortion is not as critical a parameter for the return path as it is for the forward path. Signal levels tend to be lower in the return path, and the bandwidth is limited, so the total power is small compared with a 77–channel forward signal.

At this point you might ask why there is a pad location at the input of the return path of an amplifier, and whether such pads ever serve a purpose? The pad is there mainly for convenience and versatility. It is useful for measuring or injecting signals if there is no directional coupler test point. It can also be used for plug-in thermal compensators or for ingress control switches (which are described in Chapter 13). Moreover, some designers have advocated the use of input pads as a method of reducing ingress. The basic idea is to run the set-top boxes or other terminal equipment as high as possible in order to get above ingress. Since these levels will be very high, a pad is needed before the return path gain stage of the amplifier. When such an input pad is used, the plant is still set up and aligned in the same way. The only difference is that every return path amplifier in the cascade will need to overcome the additional loss of the input pad in the station it feeds. But with the method described earlier in the chapter, the terminal equipment is already running at the highest levels possible, so input padding is unnecessary. Recall also that hybrid return amplifiers are not overdriven by signal powers that can reasonably be expected in the plant.[11]

This leads us to a more general statement that illustrates one of the most significant differences between forward and return path alignments:

Forward amplifiers are padded and equalized prior to their output hybrids, while return amplifiers are padded and equalized after their hybrids.

In the forward path, we equalize and pad at the input of the station to compensate for the cable and other loss preceding the station. We do this because there is only one path feeding the input of its amplifier from its upstream amplifier. The forward path output, on the other hand, can feed multiple paths to multiple stations. It would be impossible to adjust the output power from this station appropriately for each of these paths. Figure 11.3 shows a common example. Amp 3 feeds both Amp 4 and Amp 5. The path losses to both Amp 4 and Amp 5 cannot be made equal by adjusting Amp 3's output. However, both Amps 4 and 5 each have only one amplifier feeding them, so their inputs can be padded appropriately.

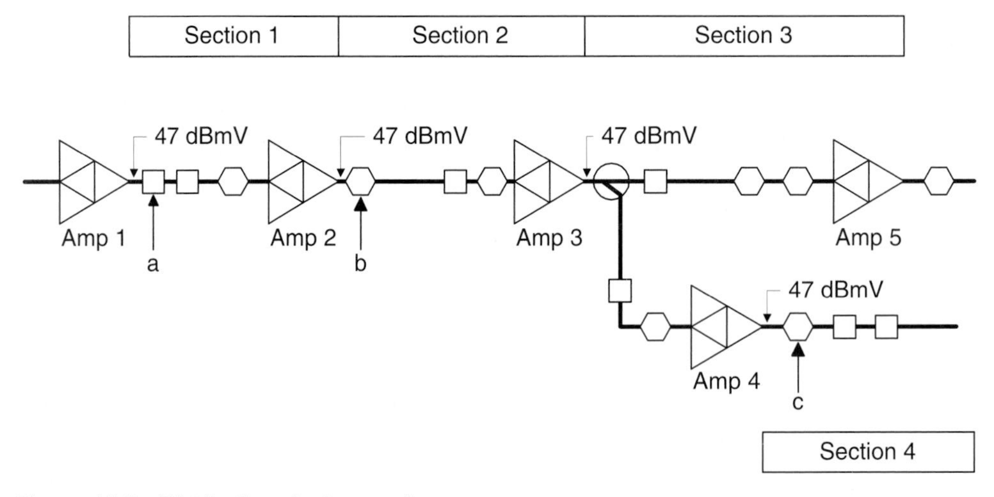

Figure 11.3 Distribution plant example

In the return path the situation is the exact opposite. Each amplifier will drive only one path toward the headend. Yet multiple amplifiers might feed into a particular amplifier. Therefore, it is not possible to compensate for the paths

11. This statement is not true for discrete amplifiers, however. See Sniezko and Werner, reference in Chapter Ten.

that feed into an amplifier. Instead the amplifier must compensate for the path that it feeds into. We therefore use padding and equalization after the return amplifier module. Referring again to the example in Figure 11.3, if Amp 3 is being aligned, it is not possible to simultaneously correct for the loss and equalization of the paths from Amp 4 and from Amp 5. Yet it is possible to correct for the loss and equalization of the path from the output of Amp 3 to the input of Amp 2. Similarly, both Amp 4 and Amp 5 can independently correct for their own paths toward Amp 3 by padding and equalizing at their outputs.

Appendices B2 and B3 discuss two special design cases that you may need to consider: (a) sections of the plant where the forward signals are operated unusually high or low and (b) trunk/bridger style amplifiers where the forward output levels for trunk ports are significantly different from the bridger ports. Other than those topics, we have completed the design discussion. It is now time to go out and align the plant!

Procedures for Aligning the Return Plant

We suggest that the plant be set up by starting at a node and working outward to the system extremities. The basic concept is that at each point in the system, a signal of the predetermined power is injected into the return input port, and the unit is padded and equalized so that the power received at the headend is at the desired level. This can be done by one person if the headend is equipped with a device to take the received signal from the return band and "echo" it back out into the system on an unused channel in the forward band (Figure 11.4). This approach, sometimes called a "round-robin," has the additional advantage of allowing both the forward and the reverse plant to be aligned during one visit to each station.[12] It can be done in either of two ways:

12. Although the plant could be set up starting from the far ends and working toward the node, it is usually easier and more straightforward to start at the node. The one advantage claimed for starting at the ends is that multiple crews can work simultaneously on the distribution from one node. In that method, each crew would work on a station and the one upstream. One person monitors reverse band signals at the upstream station and radios information back to the person padding the station that is being aligned. This, of course, requires more people and more equipment. In the method we are describing, once the node is aligned, multiple people can work on different branches from that node simultaneously. This is provided for in many of the sweep systems currently on the market.

1. The signals arriving at the headend from the return plant can be monitored by a spectrum analyzer. The display of the spectrum analyzer can then be placed on a forward path channel with a forward path modulator. The signal for the forward path can be obtained by using the composite video output connector on the spectrum analyzer, if one is available, or by aiming a TV camera at the spectrum analyzer display, as shown in Figure 11.4. The field technician injects the signals in the plant with a tone generator and looks at the spectrum analyzer's output with a portable TV set. When this method is used, one must be careful to use enough tones so that the frequency response of the return system can be accurately monitored.[13] At least four tones are required for a 35 MHz bandpass.
2. The signals arriving at the headend can be analyzed by a sweep system. Several manufacturers make systems that consist of a hand-held unit for the field technician and a rack-mount unit for the headend. The unit in the field puts out signals that are received in the headend. The unit in the headend analyzes the signals and sends the display information out into the forward path in a narrow band digital signal. This digital signal is then detected by the field unit and displayed.[14]

We need to take a moment to define our convention for naming the station ports.

We will use forward path nomenclature when referring to the station ports. This means that when we say "output port" we are using the forward path sense—this will be the return path input port.

Since the forward port names are so well established on station labels and in training manuals and other documents, we think it is less confusing to retain

13. Remember: Total power (in dBmV) = dBmV per tone + 10*log(number of tones). Examples: for four tones, total power = tone level + 6. For ten tones, total power = tone level + 10.

14. There are two key advantages to this method. The first is that the sweep provides higher frequency resolution than the multi-tone approach, thus ensuring that response "suck-outs" will not go undetected. Second, the information sent back on the forward path does not require a full 6 MHz channel.

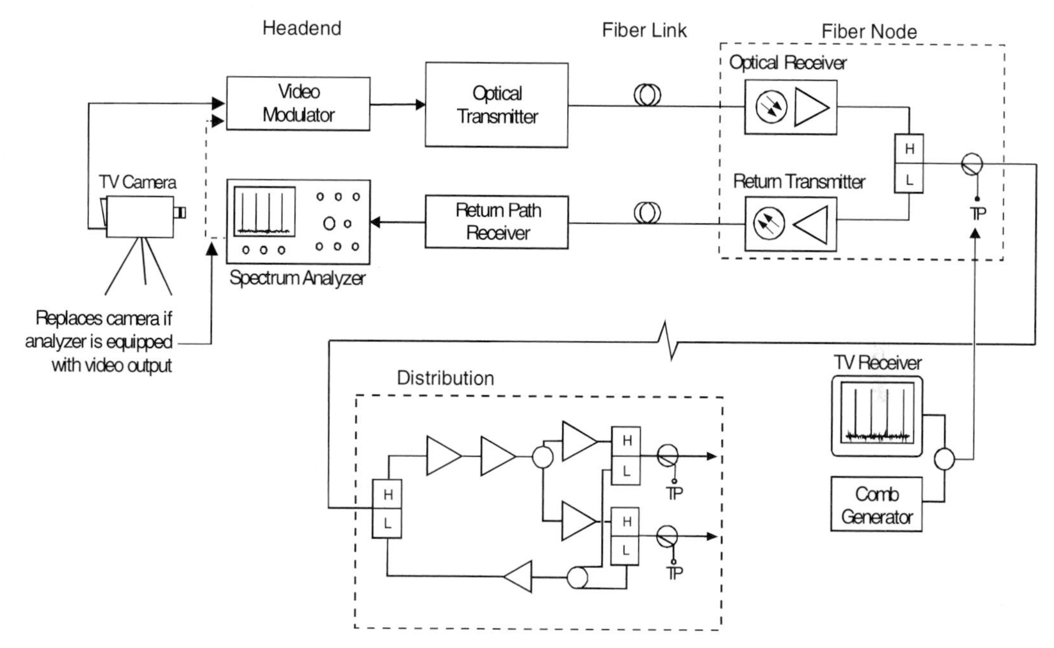

Figure 11.4 Round-robin setup

this seemingly wrong-ways view. We hope the reader will accept and understand our usage when our instructions say things like "Apply the signal to the output port."

Regardless of whether a sweep system or a spectrum analyzer is used, the procedure for aligning the return path follows the same basic outline:

1. The optical links and their receivers are first set for unity gain at the headend.
2. The field technician then starts by inserting a signal into the node station. Usually, the first injection point will be directly into the laser transmitter in order to verify the optical link gain.
3. The injection is then moved to the node output port in order to set the node return gain (as calculated earlier in this chapter). If the node return gain is not correct, it is adjusted by adding gain or loss between the station port and the laser transmitter.
4. Once the node return gain is correctly set, the gain reading is memorized as the desired *unity gain reference*.

5. The field technician then moves out in the cascade, one amplifier station at a time. At each station, he or she injects into the station's output port the same power as was applied at the node station's output port. The amplifier station's return output pad and equalizer are adjusted to come as close as possible to the memorized desired unity gain power.

This process will be described in more detail in the following sections.

Should we use "signal powers" or "gains and losses"?

Two different views can be used while aligning the return path. One can think either in terms of signal power or of gains and losses. In describing the setup procedure, we will show the relative merits of each approach and will focus on whichever method is more useful at the time. In general, however, the gain/loss approach seems to be easier to use and understand. This is due mainly to the fact that modern sweep systems are designed to provide normalized gain sweeps and are not as well suited to absolute signal level measurements.

Achieving unity gain in the optical links

Recall that when we were choosing operating powers, we needed to minimize level variations due to optical link length (Table 11-2). That means that to the extent possible, all optical links will need to have the same gain. It is not sufficient merely to have the laser transmitters in all of the nodes driven by the correct RF powers; we need also to make sure that all links have the same gain to the headend.

The first step is to determine which link will have the lowest levels. If all the lasers used in the plant have the same power, this will be the link with the most optical loss. We use this link as the reference—even if the complete alignment of this particular link can't be done first because of factors such as inconvenient location or other priorities.

Once the link with the lowest levels has been found, it will be used as the basis for determining the desired link gain. All the other links, which will have higher net gains, will be padded down to equal the long link. The padding is done either by an internal gain control in the receiver unit or by an external in-line pad on the receiver output. It should be noted, however, that in some cases there may be a couple of links that are much longer than the vast majority of links. It would be inefficient to use rare, very low gain links as the reference. A more representative link should be chosen as the reference, and post-amplifier gain should be added to the very long links to bring them up to that reference. In some cases it may even make sense to classify the links into a small number of length groups and to establish the appropriate levels—per the procedure to be described—separately for each group.

For illustration we list in Table 11-5 the nominal output for our company's return receivers when used with the laser transmitters shown in Table 11-1. By "nominal" we mean mid-range of the internal manual gain control. The actual output can be set to nominal plus or minus 8 dB.

Table 11-5 Nominal Receiver Output Power (dBmV)

Optical Link Loss (dB)	AM-RPTD	AM-RPTV1	AM-RPTV4	AM-MB-RPTD	AM-BTN-RPTV1	AM-TC-RPT	SG2-DFBT	SG2-IFPT
0	48	48	61	48	48	48	51	48
1	46	46	59	46	46	46	49	46
2	44	44	57	44	44	44	47	44
3	42	42	55	42	42	42	45	42
4	40	40	53	40	40	40	43	40
5	38	38	51	38	38	38	41	38
6	36	36	49	36	36	36	39	36
7	34	34	47	34	34	34	37	34
8	32	32	45	32	32	32	35	32
9	30	30	43	30	30	30	33	30
10	28	28	41	28	28	28	31	28
11	26	26	39	26	26	26	29	26
12	24	24	37	24	24	24	27	24

It is important to check the proper operating ranges for the receivers. For instance, the receivers referred to in Table 11-5 have best distortion performance when operated below 40 dBmV total output power. Performance is acceptable, but starts to degrade between 40 and 50 dBmV. The receivers should not be operated above 50 dBmV output. Thus it is probably a good practice to shade or color-code the cells of the table that are approaching or exceeding the operating limits, as we have done in Table 11-5. The gain of the receiver will have to be reduced in these short-link cases.

As mentioned previously, one can think of aligning the plant in terms of powers or gains. Powers are useful because they relate to the services that will occupy the return path. Gains are useful because the sweep equipment is designed to show gains. We can readily convert the power information of Table 11-5 into gains by subtracting the transmitter module input powers given in Table 11-1. This is done in Table 11-6.

Since the gain table does not give an indication of how high the signal powers are, the output power chart must be consulted in order to be sure that the fiberoptic receiver is not being overdriven.

Signal injection in amplifiers—using injection tables

Let's examine in detail where to inject signals into the amplifier and node stations. We previously explained why the unity gain point for return path alignment should be at the output port (or, more specifically, at the diplex filter).[15] Signal injection would be easy if all stations had an insertion point at the diplex filter, but this is not always the case. Fortunately, it is not hard to correct for the loss between the available insertion point and the diplex filter. For instance, consider a four-port amplifier for which a simplified block diagram is shown in Figure 11.5.

Let's say that our design calls for a particular application (set-top converter return, for example) to operate at a level of 10 dBmV at the diplex filter. For the station in Figure 11.5, this would result in a 1 dBmV signal at the input to the return path hybrid amplifier. The only insertion point is through a 16 dB direc-

15. See Appendix B1.

Table 11-6 Nominal Optical Link Gain—Transmitter Input to Receiver Output (dB)

Optical Link Loss (dB)	AM-RPTD	AM-RPTV1	AM-RPTV4	AM-MB-RPTD	AM-BTN-RPTV1	AM-TC-RPT	SG2-DFBT	SG2-IFPT
0	28	23	36	8	3	3	36	33
1	26	21	34	6	1	1	34	31
2	24	19	32	4	–1	–1	32	29
3	22	17	30	2	–3	–3	30	27
4	20	15	28	0	–5	–5	28	25
5	18	13	26	–2	–7	–7	26	23
6	16	11	24	–4	–9	–9	24	21
7	14	9	22	–6	–11	–11	22	19
8	12	7	20	–8	–13	–13	20	17
9	10	5	18	–10	–15	–15	18	15
10	8	3	16	–12	–17	–17	16	13
11	6	1	14	–14	–19	–19	14	11
12	4	–1	12	–16	–21	–21	12	9

tional coupler at the input of the hybrid. We need to calculate what signal power should to be injected at that point to be equivalent to 10 dBmV injected at the diplexer. The answer is 17 dBmV, since that power, when dropped through the 16 dB directional coupler loss, will produce 1 dBmV at the hybrid.

It is not necessary to go through these calculations every time an amplifier is aligned. A table can be established with the help of the station equipment vendor that will provide all the necessary details about where the best insertion point is and what the injection power should be. Table 11-7 gives an example for our company's products.

Don't be alarmed that some of the "insertion point loss" numbers are negative. In these instances, there is a net gain between the insertion power and the

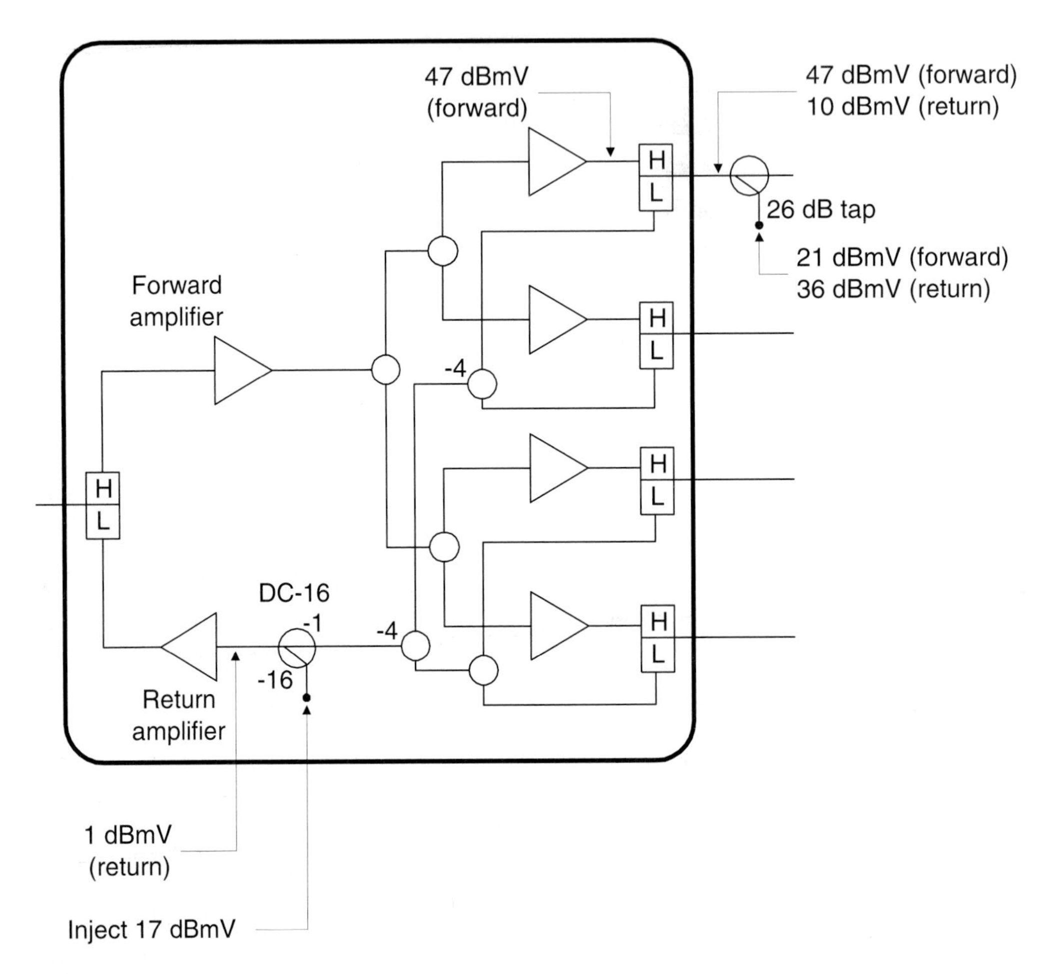

Figure 11.5 Simplified block diagram of four-port amplifier

equivalent power at the diplex filter. Note also that some amplifiers offer a choice of insertion points.

The choice of a +10 dBmV in the above example was arbitrary. For actual plant setup, one should work with levels that are a good compromise between high levels (providing good C/N) and low levels (interfering less in an operating plant), with the total power equal to that chosen by the method illustrated in Table 11-4. A sweep system can be set to equal the total power, while a multi-tone injection should be set so that each tone is 10*log (number of tones) below the desired total power.

Table 11-7 Injection Table Example

Product	Number of output ports	Test point (TP)	Directional coupler loss at TP (dB)	Loss between diplex filter and TP (dB)	Insertion point loss (IP) (dB)
			Column 4	Column 5	= Col 4 – Col 5
NODES					
SG2	4	Output test point	20	0	20
BTN	4	Status monitor transmitter	16	9	7
MBR	2	Output test point	20	0	20
MBR	2	Pad before return hybrid	0	4	–4
SX	2–4	Return test point	20	1	19
SX	2–4	Feedermaker test point	30	0	30
AMPS					
BTD-86	4	Output test point	20	0	20
MB-86	2	Output test point	20	0	20
BTD	4	Status monitor transmitter	16	9	7
MB	2	Output test point	20	0	20
MB	2	Pad before return hybrid	0	4	–4
BLE	1	Test point before return hybrid	20	0	20
BLE	1	Pad before return hybrid	0	1	–1
JLX	1	Output test point	30	0	30
JLX	1	Pad before return hybrid	0	1	–1

If a sweep system is being used, then the "injection point loss" such as that given in Table 11-7 can be used in several ways:

1. Some sweep systems allow the user to tell the system how much insertion point loss there is. In this case, the insertion point loss number from the table can be entered directly.
2. If the sweep system does not allow the insertion point loss to be entered as a correction factor, then the insertion loss will cause a lower gain to be displayed. As a result, the plant will be correctly adjusted to unity gain when the sweep gain is below the unity gain reference by the same amount as the insertion point loss.
3. IMPORTANT: In the case where the sweep system does not allow the insertion loss to be entered as a correction factor, care must be used when defining the unity gain reference at the node station. In that case, the reference at the node station must be set higher than the actual sweep reading by an amount equal to the insertion point loss given in the table. Notice that this is simply the reverse way of stating the concept given in item 2, above.

If tones are being used instead of a sweep, then the tones should be raised by the amount of the insertion point loss given in the Injection Table before inserting them into the insertion point. This rule should be observed both at the node station and at all amplifiers in the plant.

Headend Distribution

So far we have succeeded in achieving constant output powers for the composite return signals at the fiberoptic receivers in the headend. In addition, the constant-power-per-Hz allocation has ensured that all applications have the same C/N at that point. Now it is time for the individual application signals to be separated and directed to the appropriate application demodulators. This will generally require several stages of splitting—and, for some applications, combining as well—and various lengths of coax cable before the signals arrive at their individual application receivers. Thus our remaining tasks are to determine the signal levels and the C/N at each application receiver.

Signal levels

The design process itself is straightforward. The objective is to provide the correct amount of gain or loss between the fiber receiver and the application receiver. In this case "correct" means that a signal at the prescribed fiber receiver output power will hit the application receiver at its desired input level. That will ensure that all of the long-loop AGCs will keep the application transmitters at the output powers we want.

The fixed fiber receiver output power is divided down by a fan-out ratio. For applications that permit combining of returns from different nodes, such as converter returns, there will be a fan-in as well. One needs to calculate whether or not this can be accomplished without in-line amplification.

We'll work through the headend design for the three applications discussed in Table 11-4. We'll use the case in which we allocated the power in 5 MHz of bandwidth to the set-top converter (even though its occupied bandwidth is only 250 kHz). We start by calculating the power density at the output of the fiber receiver. This is the total output power (let's say this is 45 dBmV—corresponding to the SG2-DFBT transmitter in Table 11-5 with an 3 dB optical loss) minus $10*\log(35*10^6)$ or –30.4 dBmV/Hz. The level for each application at the fiber receiver output is equal to the power density plus 10*log(bandwidth allocated to the application), which is 1 MHz for the cable modems, 5 MHz for the set-tops and 6 MHz for PCS (see Table 11-8). In order to separate the individual applications, the composite signal is passed through a fan-out splitter (we assume an 8-way to allow for future service applications), which reduces all levels by approximately 11 dB plus cable loss. The set-top converter signals from eight nodes will be combined, which entails another 11 dB hit to the set-top levels.

We can see in these examples that the signals from the cable modems and the converters arrive at their demodulators with levels below the specified level (which we have arbitrarily set at +15 dBmV for the modems and set-tops and +20 dBmV for the PCS system). This means that gain needs to be inserted into each of these application circuits.[16] This needs to be done before the signal lev-

16. For the set-top converters, the gain may need to be applied separately to each of the signals that are being combined, to make up for different cable lengths and other variations.

Table 11-8 Headend level examples

Application	Unit	Cable modem	Set-top converter	PCS	Notes
Allocated bandwidth	MHz	1	5	6	
Fiber receiver output power	dBmV	45	45	45	
Power density at fiber receiver	dBmV/Hz	–30.4	–30.4	–30.4	Subtract $10*\log(35*10^6) = 75.4$
Application level at fiber receiver	dBmV	+29.6	+36.5	+37.3	Add 10*log(allocated bandwidth)
Cable loss	dB	–2	–2	–2	
Fan-out loss	dB	–11	–11	–11	8-way
Cable loss	dB		–2		
Combining loss	dB		–11		8-way for STB, only
Cable loss	dB	–3	–2	–3	
Net level	dBmV	+13.6	+8.5	+21.3	Subtract losses
Level required at application receiver	dBmV	+15	+15	+20	From specifications for the applications
Gain required	dB	1.4	6.5	–1 (pad)	

els fall below 5 dBmV, so that the C/N is not degraded unnecessarily. For wide-band signals, there might also be some correction needed for cable tilt, as well.

Carrier to noise

Our last task is to see that the carrier to noise is adequate for each application. In concept, all of the applications emerge from the fiber receiver with the same C/N, as a direct outcome of the constant-power-per-Hz method. Recall, however, that in our example we allowed the set-tops to operate at higher levels by allocating to them all of the power associated with the 5–10 MHz part of the return band. That means that the signal levels for the set-tops will have been boosted by 10*log (5 MHz/250 kHz), or 13 dB, which will raise the C/N by a similar amount.

If we were calculating C/N for forward path signals that we were putting through splitters, amplifiers, and combiners, we would have to be concerned about degradations at each step. For the return signals in the headend, only the combining stages will cause C/N decreases. An N-way combine of approximately equal C/Ns will degrade the net C/N by 10*logN, because we are adding in N more noise sources. On the other hand, if we insert amplifiers into the signal path so that signal levels always stay above about 5 dBmV, we can be sure that amplifier thermal noise and splitting losses will not cause measurable degradation.[17] The C/N for the three applications is summarized in Table 11-9.

Table 11-9 Headend carrier to noise

Application	Cable modem	Set-top	PCS	
C/N at fiber receiver	40	40	40	Typical
C/N adjustment		+13		10*log(5 MHz/250 kHz)
8-way combine		–9		STB only
C/N at application receiver	40	44	40	

Note that we have accounted only for the noise generated by the fiberoptic and electronic equipment. We need also to allow for ingress. This is particularly relevant to the converters in our example (a) because the 8-way combine will boost the single feed ingress by 9dB (assuming that each of the other feeds has

17. From Chapter Four, we have that C/N_{amp} = Input – NF – (noise floor). We expect C/Ns on the order of 40 dB out of the headend fiber receiver. From Table D-2 (Appendix D) we can see that if we combine a 40 dB C/N with a C/N≥ 53, there will be less than a 0.2 dB degradation. According to Table 4-1, for the signals expected in the return path, the noise floor will range from –75 dBmV (10 kHz bandwidth) to –57 dBmV (6 MHz bandwidth). This means that a (relatively noisy) amplifier with a noise figure of 8 dB would generate a 53 dB C/N for 6 MHz signals above 4 dBmV. Narrower-bandwidth signals could maintain C/N with even lower levels.

the same ingress level) and (b) because the set-tops may be operating in the most ingress-prone portion of the band.

Headend tips

Headend interconnections are very complex. Therefore, the return design and setup will almost necessarily be complicated and time-consuming. There are some practices that will help, however.

The first is to start with the assumption that there will be changes to the interconnections as the services evolve. This makes it essential to keep very good records of all splits, combines, cable losses, and amplifications and to put informative labels on every component and cable. A suggestion that is likely to pay off is to use only one type of splitter/combiner, say an 8-way.[18] This simplifies the design and may even allow for some "painless" future re-configurations.

Last, a word of caution. We have spent a great deal of care in designing the return system so that all of the long-loop AGCs will maintain proper operating levels in the plant. The headend engineer needs to be sure that the whole plant alignment doesn't get thrown off by someone coming along afterward and inserting an additional splitter into the headend cabling. All personnel need to be made aware of the care that has been invested in achieving the correct set-up. A training session that—at the minimum—conveys to all personnel the concept that everything is interrelated would likely be beneficial.

Summary

The discussion in this chapter has dealt with plant design and alignment in considerable detail. This is because these are probably the most critical subjects we have covered. Experienced operators all agree that the success or failure of the service applications operating over the return path depends precisely on whether the design and alignment are done well. Improperly aligned plant will not only be trouble-prone but will also be very difficult to trouble-shoot.

18. O.J. Sniezko, "Multi-Layer Headend Combining Network Design for Broadcast, Local and Targeted Services," NCTA Technical Papers, pp. 300–15 (1997).

In fact, our treatment of this process is still not detailed enough. Fortunately, detailed set-up procedures have been generated both by operators and by vendors of distribution equipment and instrumentation.[19]

For reference, we conclude with an overview table of the most significant differences between forward and return designs and setups.

Table 11-10 Key differences between forward and return path alignments

	Forward path	Return path
Primary alignment goal	Constant output with low distortion	Constant input with minimum noise
Pad and equalizer location	Before hybrids	After hybrid
Method of alignment	Measure output power at station	Inject test signal at station and measure output at headend

Summing-Up...

- The return path unity gain reference point should be the diplex filter at the return input port for both amplifier and node stations.
- RF plant levels should be set as high as possible, limited by the output capability of in-home transmitters on a power-per-hertz basis.
- Laser transmitter inputs are set to the predetermined operating point with the aid of node gain and injection information.
- Separate level and C/N determinations are needed for the distribution within the headend.

19. Return Path Level Selection, Setup, and Alignment Procedure," Reference Guide 453760-001, NextLevel Systems, Inc., 1997.

CHAPTER 12

Architectures

Up to this point we have assumed that the HFC network is a localized plant with one headend and single-span fiberoptic links feeding nodes and RF distribution. In this chapter we will broaden our view to include more extensive networks that have multiple headends and cascaded fiber links. To make the discussion understandable, we will need to describe both forward and return portions of such networks, but our focus will continue to be on the particular concerns of return transmission. Among the issues treated in this chapter will be:

- How is narrowcasting done in distributed networks?
- What are the architectural differences between forward and return paths?
- How can the return signals traverse multiple fiber spans?
- What are the options for long fiber links?

Regional Networks

It is becoming more and more common for a *multiple system operator* (MSO) to acquire ownership of several adjoining plants and to combine them into a larger *regional network*.[1] These regional networks will typically consist of one or two *master headends*, where signals are collected from satellite and land-based microwave antennas and from other sources, and a *backbone distribution* to a number of *primary distribution hubs*. The primary hubs may distribute signals directly to nodes or may redistribute them to *secondary hubs*[2] that connect to fiber nodes. In a regional system serving 500,000 homes, the primary hubs might serve 100,000 homes, the secondary hubs 20,000 homes, and the nodes 500 to 2000 homes. Since so many homes depend on the network connecting the primary hubs with the master headend, it is usual for the backbone to incorporate redundant, route-diverse connections. Often this takes the form of a bidirectional multi-fiber ring (Figure 12.1). Note that as a general rule, all of the headends and primary hubs provide direct signal distribution to nearby nodes.

It is instructive at this point to work through an estimate of fiber counts and equipment space requirements. The example of the previous paragraph implies that there are approximately five primary hubs (one of which is co-located with the main headend). Each primary hub feeds five secondary hubs, and each secondary feeds 20 nodes. At four fibers per node (forward, return, and 2 spares) there will be 80 fibers downstream from the secondary hub. Identical fiber provisioning levels between the primary and secondary hubs would imply 400 fibers into the primary, which is likely to be an unmanageably large number, so some degree of fiber concentration appears to be needed. Let's assume that there is 4:1 fiber concentration accomplished, for instance, by block conversion of forward and of return narrowcast information.[3] This reduces the total on the downstream side of each primary hub to approximately 100 fibers.

1. D. Raskin and C. Smith, "Regional Networks for Broadband Cable Television Operations, " National Fiber Optic Engineers Conference, vol 2, pp 351–60, (1997).

2. The term *optical transition node* (OTN) has been used to denote a secondary hub, especially one housed in a pedestal, rather than a building, but that term is becoming less common.

3. For example, the forward narrowcast signals in the frequency band 550–750 MHz that are destined for a given node are shifted as a group to lower frequencies and combined with 200-MHz-wide blocks for three other nodes that have been shifted to non-overlapping bands. The combined RF signal is put on a single laser for transmission from primary hub to secondary. At the secondary hub the blocks are separated, restored to the 550–750 MHz band and individually combined with a common 52–550 MHz broadcast signal for transmission by four different lasers to the appropriate nodes.

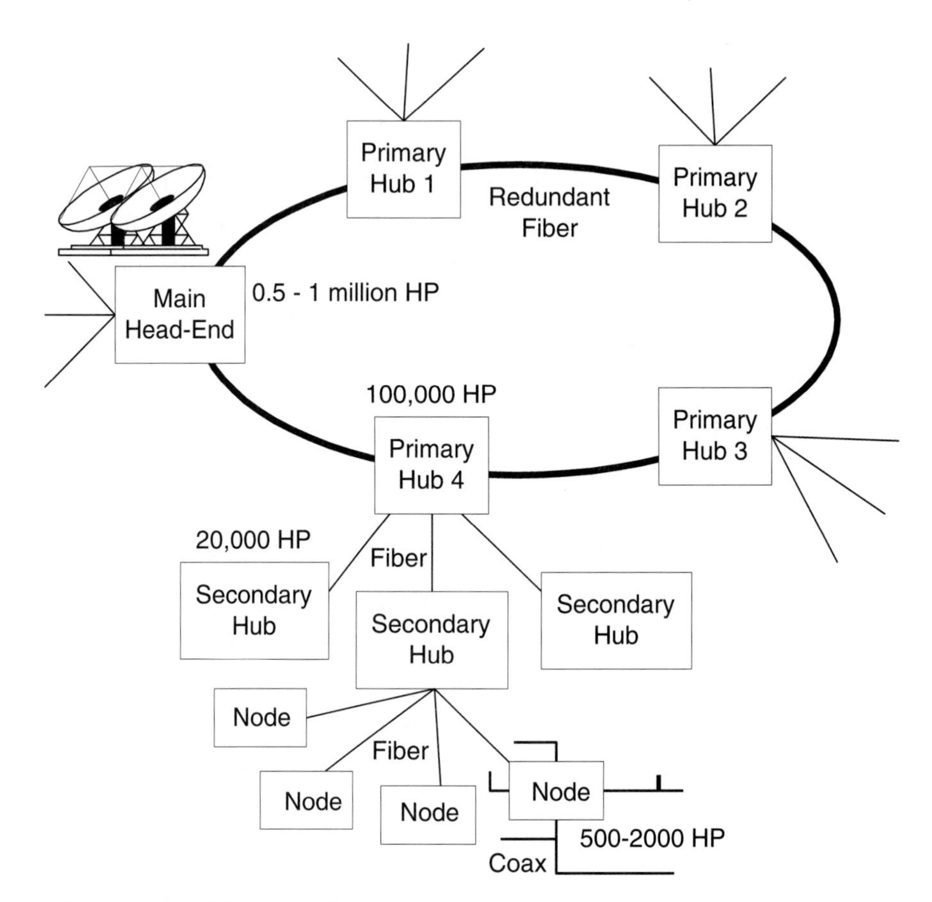

Figure 12.1 Regional ring network

The equipment required at the secondary hub is

1 forward broadcast receiver
1 forward amplifier (to fan-out to 20 laser transmitters)
5 forward receivers (@1 per 4 nodes)
5 forward block converters (@1 per 4 nodes)
20 forward lasers (@1 per node)
20 return receivers (@1 per node)
5 return block converters (@1 per 4 nodes)
5 return transmitters (@1 per 4 nodes).

Most transmission equipment vendors supply these pieces in the form of slide-in modules in a card cage or shelf arrangement. Let's assume eight modules per cage and an 8-3/4" height (5U) for the cage. The 62 modules fit into approximately eight cages, which correspond to 70" of rack height. While a higher packing density may be possible for some of the modules, space for power conditioning, back-up facility, and fans or cooling must still be added. This corresponds to a fairly large "hut" or to a small section of a building. Of course if application electronics such as cable modem termination systems (CMTS) must be located at the secondary hub, the required amount of rack space will multiply rapidly.

Analog transport backbones

If the distances are not too great and if the emphasis is on forward path broadcast services, the least costly approach to regional distribution is to use 1550 nm analog fiberoptics. Transmission at 1550 nm has two advantages over the traditional 1310 nm: the fiber attenuation is approximately 30 percent lower, and the signals can be amplified in the optical domain by devices called *erbium doped fiber amplifiers* (EDFA) without first converting back to RF electrical signals. The disadvantage of 1550 nm is that the transmitters are inherently expensive—more than ten times the cost of 1310 nm—because they must be chirp-free.[4]

Transmitters with output powers of 16 to 20 dBm (40 to 100 mW) are available commercially. With a typical fiber loss of 0.25 dB/km, it is possible to produce single spans of 60 km and—with an in-line EDFA in the outside plant—up to 100 km. Distances much beyond 100 km are limited by optical nonlinearities in the fiber, which can only be overcome by methods that are restrictive in other

4. As mentioned in Chapter Nine, laser chirp refers to the laser wavelength changing as the signal amplitude increases and decreases. For 1310 nm DFB lasers operating in standard singlemode fiber (which is designed to have zero dispersion at 1310 nm), the chirping is generally of no concern. A 1550 nm laser operating with the same chirp into 1310 nm fiber would generate severe even-order distortion of the RF signal, resulting in unacceptable CSO performance. To understand the distortion process, consider a signal consisting of a single sinewave. The positive-going signal will chirp the wavelength up, which causes a slower propagating optical wave (due to the fiber dispersion). The negative-going signal will chirp the wavelength down, which causes a faster propagating optical wave. Accordingly, the receiver detects a distorted RF wave: the peak of the positive peak is retarded from its normal position, and the negative peak is advanced. Since the distortion is symmetrical, it decomposes into even-order harmonics.

ways.[5] The reach of 1550 nm transmission can be increased beyond that limit by dividing the analog channels among multiple fibers. In cases where 1550 nm systems can deliver acceptable performance, the 1550 nm system cost is likely to be considerably less than any baseband digital alternative.

In a typical 1550 nm regional system, the optical signals are detected back to electrical RF at the primary or secondary hub. At that point the signals are amplified and fanned out to several 1310 nm lasers for transmission to the nodes. For narrowcast applications, the forward narrowcast signals are combined with the broadcast signal at the 1310 nm laser. This, of course, raises the question of how the narrowcast signals get to the hub in the first place. At this point the 1550 nm scenario begins to get more troublesome because the high cost of the transmitters is unfavorable to narrowcasting. One possibility is to transmit several bands of narrowcast signals on a single 1550 nm transmitter by use of the block conversion techniques discussed earlier in this chapter.

A costly but more "future-proof" method is to run a parallel digital-only network to the hubs for narrowcasting, using separate fibers within the backbone cable sheath. This can provide a straightforward solution as well to the next question: How do we get the *return* signals back to the main headend? We will discuss digital transport networks in the next section. For the remainder of this section we will discuss how to accomplish the return in the analog domain.[6]

5. G. Gopalakrishnan et al., "Experimental Study of Fibre Induced Distortions in Externally Modulated 1550 nm Analogue CATV Links," *IEE Electonics Letters*, vol 32, pp 1309–10, (Jul 96).

   F. Willems et al., "Experimental Verification of Self-Phase Modulation Induced Nonlinear Distortion In Externally Modulated AM-VSB Lightwave Systems," *IEE Electonics Letters*, vol 32, pp 1310–11 (Jul 96).

   C. Desem et al., "Composite Second Order Distortion Due to Self-Phase Modulation in Externally Modulated AM-SCM Systems Operating at 1550 nm," *IEE Electonics Letters*, vol 30, pp 2055-6, (Nov 94).

6. Remember that, even though the return payload generally consists of *digital* signals, these are digital signals that are modulated onto a multiplicity of carriers. Until each of the individual signals within the return spectrum is detected and demodulated to baseband by its application receiver, we need to treat the payload as if it is an analog signal. The one divergence from forward video analog is, of course, that the C/N and distortion requirements aren't as stringent.

Because of the complexity of 1550 nm transmitters, it is not feasible to fit a transmitter into a node station for the return, so we are restricted to 1310 nm technology. This means that we can't expect the "reach" of the return transmitter to be as long as that of a forward 1550 nm unit. In relatively simple systems this may not be a problem, as illustrated by the following example.

Consider a cable TV operation in which all of the services are broadcast to all subscribers, but access to some channels is limited by scrambling. Using a set-top converter, customers order pay-per-view and premium services via the return system, which also carries a status monitoring signal. For the forward system, the primary hubs are merely passive splitting points: the signals from the master headend are distributed on a 1550 nm system all the way to the secondary hubs. At the secondary hubs, the 1550 nm signal is detected and re-lased at 1310 nm to the nodes. This is a very economical network, even when multiple fibers are needed in the backbone to carry the full channel loading.

The operator wants to minimize the cost of the return system, which is relatively lightly loaded, by placing the data receivers at the few primary hubs, rather than at the much larger number of secondaries.

Let's assume that the worst case (longest) distances are:

Master headend to primary hub:	*50*	*km (12 dB @ 1550 or 20 dB @ 1310)*
Primary hub to secondary hub:	*25*	*km (6 dB @ 1550 or 10 dB @ 1310)*
Secondary hub to node:	*5*	*km[a] (2 dB @ 1310)*

In this case we can see that the forward splitting at the secondary hub works in our favor. A single fiber hop from the node to the primary hub amounts to no more than 12 dB for 1310 nm lasers. Since the return plant is lightly loaded, carrying only the IPPV and status monitoring signals, the signal levels into the return laser can be raised to ensure that the 12 dB span will be bridged.

The IPPV and status monitoring receivers at the primary hub convert the signals into their baseband digital form. The problem of getting the signals back to the main headend is thus reduced to a straightforward data communications task.[b] A number of vendors make point-to-point fiber modems that can readily span a 20 dB link. Alternatively, standard telephone modems can be used over private lines. We will discuss arrangements for remote monitoring in more detail in the next chapter.

a. We are assuming that the forward signals from the Secondary Hub to the nodes are optically split four ways.

b. Some IPPV systems provide equipment for demodulating the return data stream (after optical-to-RF detection) and then remodulating. This remodulated signal is sufficiently cleaned up that it can be applied to a laser for further transmission. In polled systems like this, equipment costs can be minimized by combining multiple signals in the RF domain, since there is no contention for the communication channel between different set-top boxes.

As can be seen from this example, extremely cost-effective two-way systems can be built with a combination of 1550 nm and 1310 nm technologies.

Another technique in the analog domain that might be useful for longer return links with heavier data loading is to use higher-power lasers. The cost of the lasers can be offset by using block conversion techniques to carry multiple returns on one laser. For an 800 MHz laser, six to eight 35 MHz return bands could be combined into a single RF payload.

It should also be apparent that as the two-way applications get more ambitious and as the network expands, the limitations of analog techniques for distributed networks will become increasingly burdensome. Thus the aim of these methods is primarily to defer larger investments until the market "pull" for highly interactive applications becomes established.

Digital transport backbones

HFC networks, using modulated carrier digital transport for narrowcasting, are extremely efficient at widespread delivery of broadband information over relatively short distances to and from large numbers of end-users. In contrast, baseband digital transport networks, as typified by inter-exchange telephone transmission, are notable for their long distance and high data rate capabilities. As the demand for highly interactive services increases, it is easy to see that there will be a point at which a regional cable network will evolve toward a hybrid of the two technologies. This is done by concentrating at hub points the relatively low-rate data flows from multiple nodes and using baseband digital transport for the aggregated data streams between these hubs and the main headend.

In order to assemble the many low-rate data flows into a single high-rate data stream, the modulated carrier signals transported from the nodes need to be RF detected, distributed to application receivers, and demodulated to baseband before being put onto the backbone ring (Figure 12.2). Since this entails a considerable amount of equipment for each application, it is preferable that this take place at the primary hubs, rather than at the more numerous secondary hubs. Of course that efficiency is possible only if the analog-based signals from the nodes can be brought back as far as the primary hub. Alternatively, it may be practical to arrange the system with the application receivers in the primary hub initially, when interactive traffic is relatively light, and then move them out to the secondary hubs as market penetration builds.

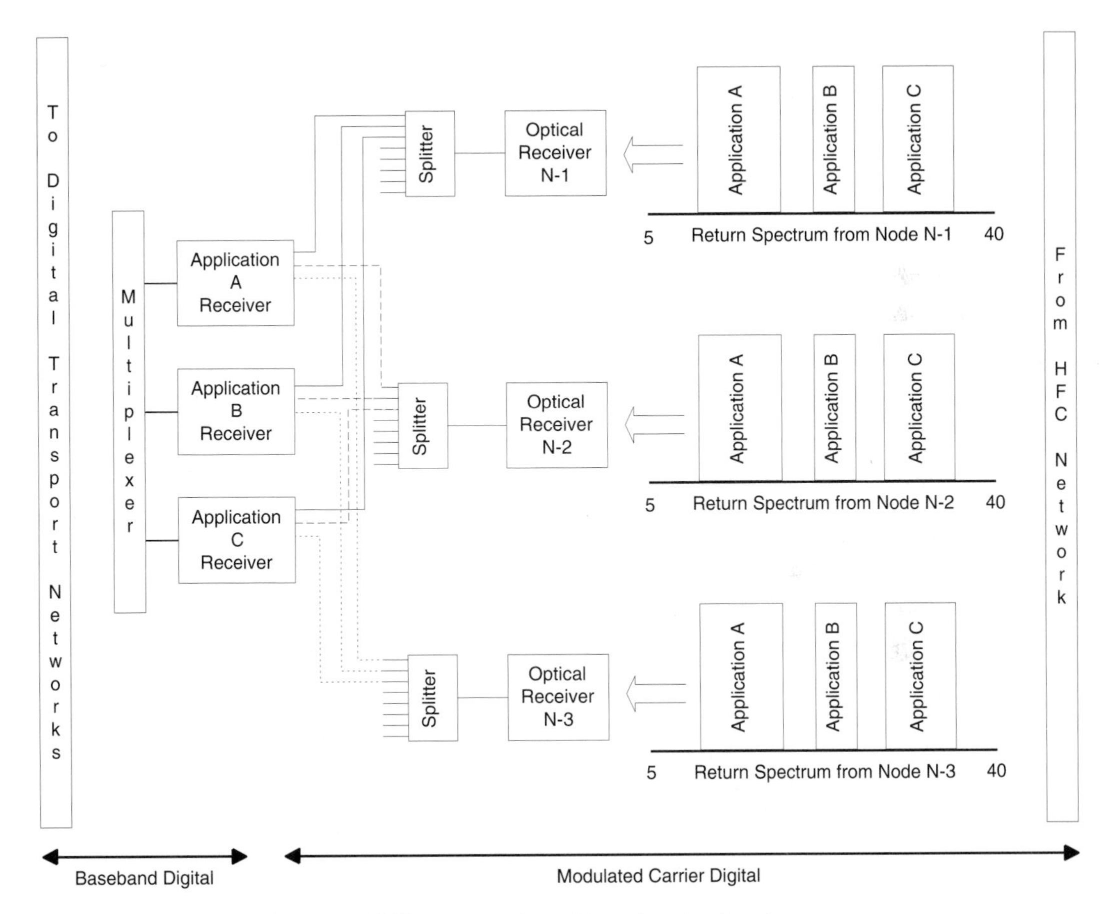

Figure 12.2 Interface between HFC return path and baseband network

Figure 12.3 shows the applications A, B, and C in more detail. For illustration the applications are chosen to be set-top box (IPPV) returns, status monitoring telemetry, and cable modems. The RF output from each optical receiver is split and sent to the appropriate bank of application receivers. For IPPV, each signal is sent to a *return path demodulator* (RPD), where it is filtered and downconverted to baseband. The output of the RPD is generally compatible with Ethernet (10Base-T), a standard data interface for bidirectional communication with the access system controller. In a regional cable network, the access controller would be located in the main headend, with the RPDs in the primary hubs. Telemetry applications such as status monitoring use Ethernet or other standard data protocol on the output of the application receiver. The cable modem termination system (CMTS) has a higher-rate data interface on its output, such as Gigabit Ethernet or Fast Ethernet. Each of these signals is fed to an appropriate interface card on a multiplexer for transport to the main headend, where the signals are demultiplexed and sent to access controller, network monitoring controller, servers, and routers.

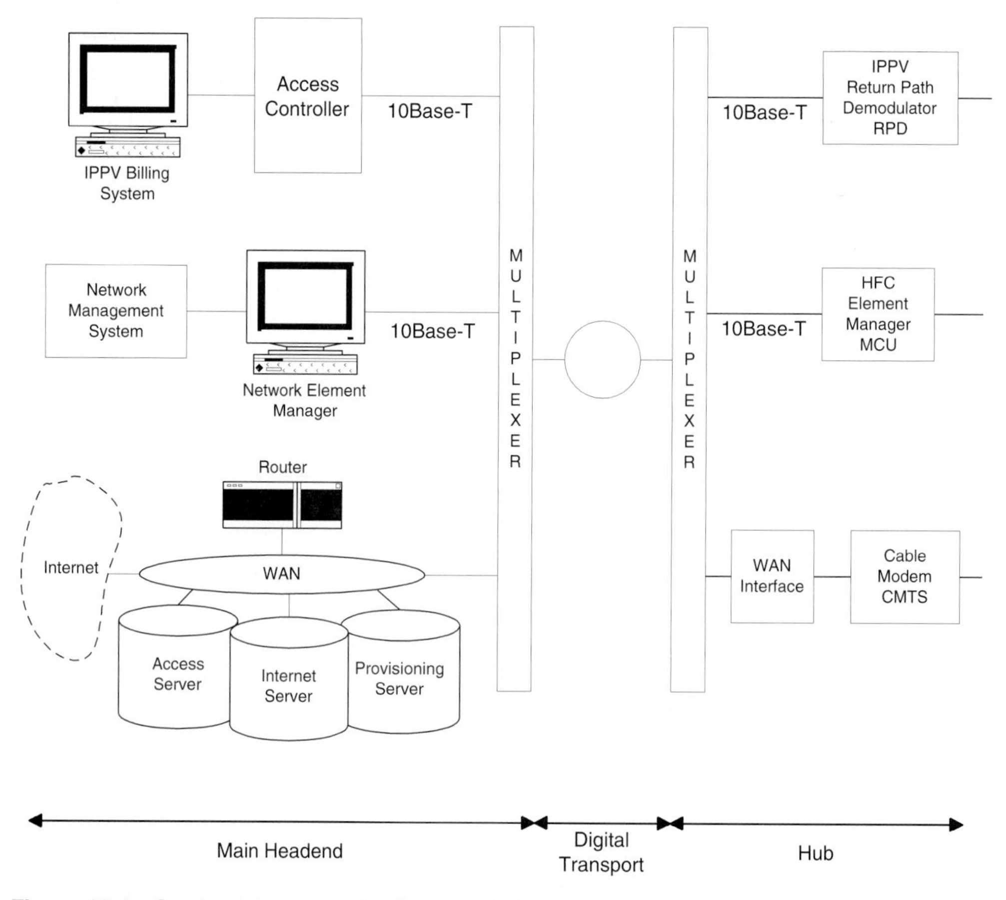

Figure 12.3 Service interconnection between hub and main headend

Digital transport standards

In Appendix C we discuss the principal high-speed digital transport standards: SONET (used in the US and Canada and in a modified form in Japan) and SDH (used elsewhere, including Europe, Asia, South and Central America, and Australia). Although a cable operator is free to choose a digital transport system that is not standards-based, the need to connect important applications, such as cable modems, to the public switched telephone network (PSTN) essentially requires standards compatibility.

The cost of the standard equipment is dropping rapidly as its deployment spreads worldwide. This means that both the transport equipment itself, the interface equipment (for getting services on and off the transport ring), and the management software are likely to be lower than proprietary alternatives. The savings associated with using a single network management system for all transport—including downstream trunking of video in digital streams—cannot be overlooked, as well. Since management software and training represent a multimillion dollar investment, it is not likely to be cost-effective to have multiple systems. (Nor is it likely to be operationally efficient.)

It is also possible to see ahead to a time when even video services will be delivered either from the PSTN or the Internet. That could make obsolete an investment in proprietary digital transport equipment.

Equipment List for an Example Network

This section is intended to give an order-of-magnitude estimate of the equipment requirements for a network that should be reasonably representative of future deployments.

Assumptions

In order to get started, we need to make some assumptions about the physical architecture of the network and about the broadcast and narrowcast services that will be flowing over it.

Architecture

We assume a regional network similar to that shown in Figure 12.1, consisting of

- a main headend serving 400,000 homes passed (HP),
- three primary hubs, each serving 100,000 HP (in addition to the 100,000 HP served directly from the main headend),
- three secondary hubs off each primary hub, each serving 24,000 HP, and
- forty-eight nodes off each secondary hub, with 500 HP per node.

The main headend and primary hubs are linked by a multiple-fiber redundant ring. All of the connections to the outside world are at the main headend, including satellite downlinks, connections to the public telephone network and the Internet, off-air antennas, and primary servers.

Services

We assume that the forward bandwidth to each node is 750 MHz and that there will be broadcast services emanating from the main headend that consist of 80 analog channels and 14 digital multiplexes, which may include near-video-on-demand. This leaves 134 MHz free to be used for downstream digital narrowcasting to each node.[7] The return bandwidth is 35 MHz.

In addition to the broadcast services, we assume a well-developed cable modem service, along with addressable analog/digital converters and network management. As we will see, there will be adequate digital bandwidth for later addition of a video-on-demand service.

If we consider one of the 500-home nodes, let's assume a 25 percent overall penetration rate for cable modems, and let's see what happens if users are able to access 1 Mbps rates in both directions. If we size the system on the basis that no more than 10 percent of the cable modem subscribers will demand their maximum rate at any one time, then the maximum bit rate in each direction for one node is 500 x 0.25 x 0.10 x 1 = 12.5 Mbps. For 16-QAM modulation in the return path (4 bits/Hz, theoretical), this would require about 4 MHz. For QPSK (2 bits/Hz, theoretical), 7 MHz would be required. More realistically, the spectral efficiency will be reduced by error correction bits, so the cable modem service could require 5–8 MHz. It is clear that with these traffic assumptions, the HFC system has abundant data capacity.

7. We are not specifically allocating spectrum to FM radio broadcasting or digital music, but that could be considered a part of the 80 analog/14 digital channel complement.

The data flowing to or from all of the 48 nodes that are connected to a secondary hub amounts to $48 \times 12.5 = 600$ Mbps.

Transporting the signals

Under reasonable assumptions about the geographic size of the plant, the forward broadcast signals will need to be transported digitally from the main headend to the primary hubs. Most likely, transporting the 94 channels will require more than one SONET OC-48 (Appendix C1) since a single OC-48 can carry 16 to 96 channels, depending on the specific equipment and channel characteristics. Transport from the primary to secondary hubs should be possible with 1550 nm techniques.

Since each secondary hub generates and receives 600 Mbps of data, an OC-48 ring connecting the hubs seems appropriate. Each secondary hub would then be capable of picking off an OC-12 stream (622 Mbps) from the OC-48. The 622 Mbps streams would be aggregated in a router in the secondary hub, as shown in Figure 12.4. The router is connected to the cable modem headend units (CMTS) by Fast Ethernet.

If the primary hub contains caching servers,[8] then it may not be necessary to transport the full OC-48 stream from each primary back to the main headend. In that case, a total of three OC-48s running on the primary ring may be sufficient to carry the broadcast and data services.

Other data services, such as conditional access and network monitoring, will interface with the SONET transport via appropriate interface cards. These services do not add much data to the overall load.

In the secondary hub the equipment for broadcast services alone will require three to four full racks. The space required for the data equipment will depend on the particular designs. Transport will require about 35 percent to 50 percent of a rack. Negative 48 volt powering is necessary at minimum for the transport equipment.

8. It is common for *Internet Service Providers* (ISP) to maintain copies of many of the popular Web sites on local servers. This improves access time for their customers by keeping their communications local.

Table 12-1 Equipment lists

Equipment lists	Primary hub	Secondary hub
Broadcast services:		
Satellite receivers	(1) to (3) 1550 nm transmitters	1550 nm receiver
Modulators	Modulators	(48) laser transmitters
Upconverters	Upconverters	(24) dual receivers
Processors	Conditional access controllers	Amplifiers
Transcoders	Local ad insertion	Combiners
Scramblers		Fiber management
Stereo encoders		Conditional access network interface
Conditional access controllers		Status monitoring demodulator/controller
Ad insertion		
Data services:		
Servers	Servers	Router
Routers	Routers	CMTSs
ATM switch		(48) Upconverters
Transport:		
(3) OC-48 multiplexers	(2) OC-48 terminals	(1) OC-48 add-drop multiplexer with OC-12 ports
Video encoders	(1) OC-48 multiplexer	
Network management domain manager	Video decoders	

Note: The lists for main headend and primary hub do not include the equipment needed for local distribution.

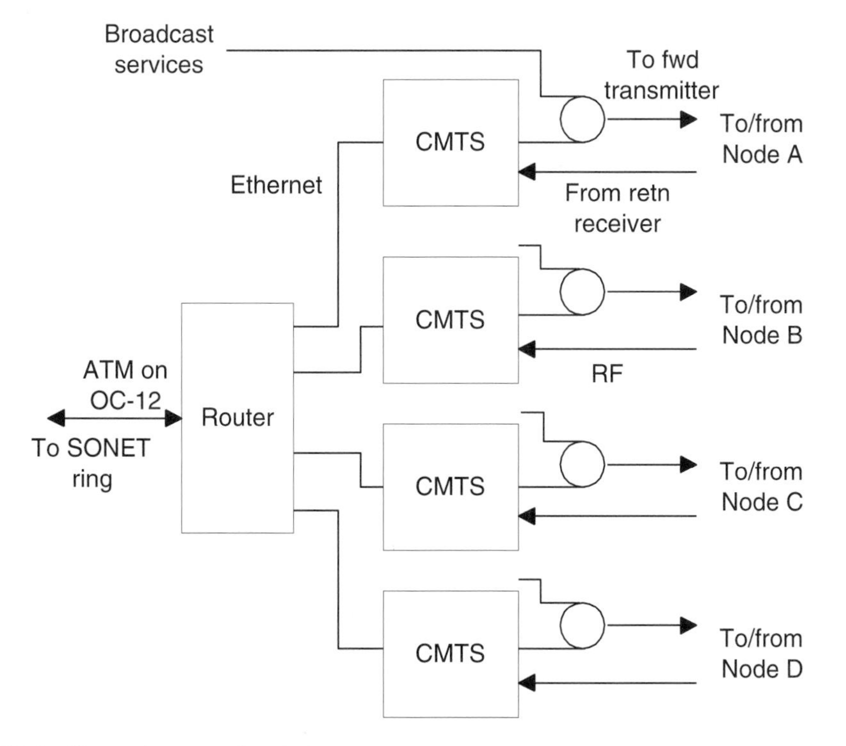

Figure 12.4 Router in secondary hub

Summing-Up...

- Regional networks are becoming prevalent.
- Analog technology using 1550 nm transmission and optical amplifiers is an economic choice when distances are not too great.
- Two-way services make analog techniques increasingly difficult to apply.
- Standards-based digital backbone distribution of video, voice, and data appears ready to become the technology of choice.

CHAPTER 13

Network Management

The broadband communication system that offers an array of two-way interactive services is a much more sophisticated one than the traditional "cable TV" system that has offered only unidirectional broadcasting of entertainment video. As we have already noted, at the same time that today's cable operator is working to master the increased complexity of these networks, today's customer is raising the service quality expectations for the network. Thus there are at least two drivers toward systems for managing the cable network:

- to improve network availability, through monitoring and pro-active maintenance, and
- to improve operation of the services running over the network by passing performance information between the HFC plant and the applications running over it.

These needs go well beyond the "status monitoring" function that has been employed in a small number of systems over the years. This chapter discusses this broadened scope by describing the network management task, establishing a standard management architecture, and discussing the hardware and software options for meeting the challenge. New directions will be discussed as well. This chapter describes many systems that are still under development.

Where Is Network Management Needed?

We start by discussing the two key objectives of a network management system: to increase the availability of the network and to coordinate the operation of the network with the performance of applications running on it.

Improving system availability

For years, it has been a running joke both within and without the cable TV industry that cable systems are monitored by their customers: when the system fails, the phones start ringing.[1] The follow-on to the joke is that the operators will need to come up with a better scheme before they offer HFC telephony.

As discussed in Chapter Seven, system unavailability is equal to the mean time between failures (MTBF) multiplied by the mean time to repair (MTTR). Network management techniques improve system availability primarily by reducing MTTR:

- by isolating the point of system failure (Distribution amplifier at 5th and Elm has failed),
- by providing data on the type of failure (Low forward output on one branch),
- by suggesting a remedy (Replace amplifier module), and
- by dispatching a technician equipped with the appropriate module.

1. The ultimate example of this is one system that regionalized its plant and provided route diversity on the primary fiber links, bringing two fibers to each primary hub. An A-B switch in the node connected the RF distribution to the primary fiber, but switched to the backup upon signal loss from the primary. Because they had not provided remote monitoring of the node, the operators had to find a way to determine when the system was using the backup fiber. Their solution was to eliminate one of the home shopping channels from the B feed. Once the system was in service, they found that they never had to wait more than a few minutes for one of their subscribers to let them know that the system had changed over to the backup fiber.

It is clear that the management system that performs these functions has fault isolation and analysis capabilities and has real-time communication with the workforce management system. In addition the management system logs all field failures in a database. These records can be analyzed off-line to find and document repetitive equipment failures. That information can be used to improve the equipment design, thus reducing MTBF as well.

Ideally the management system will do trend analysis to pinpoint incipient equipment failures, so these get fixed before they disrupt service.

Managing interactions between services and the physical plant

Because it contains remotely controllable elements, today's physical plant *needs* to be managed. Many nodes include *ingress control switches* (ICS) that can be commanded via a status monitoring transponder in the node to insert either a small (~ 6 dB) or a large (~ 40 dB) amount of attenuation into a given return input leg. The purpose of an ICS is to help isolate return ingress problems. If switching-in 6 dB of padding reduces a troublesome noise level by nearly 6 dB, then the operator can be quite sure that the noise is entering the system through that leg.[2] For a system with hundreds of nodes, each with dozens of amplifiers, this method of ingress localization can be painfully time-consuming. A better way would be for each application receiver to check the performance of the communication channels feeding it and to report BER information to the management system. Automated searches to pinpoint ingress sources could then be done under the direction of the manager.

In addition, if an ICS is set to attenuate, the headend equipment controlling in-home transmitters, such as two-way set-top boxes and modems, needs to be informed. Otherwise a downstream poll will not be acknowledged, which would cause the controller to command the transmitter to raise its output. Ultimately this could result in a number of in-home transmitters being set to "scream." Sub-

2. A 40 dB pad essentially opens the input connection, but a 6 dB pad may allow certain types of data traffic through without loss. Thus a 6 dB ICS would allow the plant to be probed for ingress "hot spots" without disrupting these data services.

sequently, when the ICS removed attenuation, upstream communication could be completely disabled by the resulting overload.

In Chapter Eleven we discussed using specific signals from in-home application transmitters such as cable modems to monitor the performance of the plant. In order to make use of this information, there needs to be coordination between the application receivers in the headend and the plant management system.

If addressable taps (Chapter Five) are deployed to control access and to keep unwanted signals from entering the return plant, their actuation needs to be controlled by an integrated system that communicates with the billing system, the HFC system, and all application controllers.

Element Management and Integrated Management Systems

We have just described several situations that call for a management system that gathers information from many different parts of the operation, analyzes and coordinates it, and does intelligent things with it. This is a major departure from the traditional hands-on operating mode of cable TV, but it is the clear direction necessitated by the increasing complexity of broadband cable networks. There is a framework for discussing such an *integrated network management system* (INMS) and the structure of other network parts that it interacts with.

Management hierarchy

The various parts of the network—the HFC plant, the cable modem system, the addressable access system, etc.—are called *network elements* (NE). Each element is controlled by its own *element manager* (EM). Thus the status-monitoring transponders built into amplifiers and node stations communicate with an HFC element manager, the cable modems communicate with a data EM, and the set-top boxes communicate with an access EM. In the operational scenarios we just described, it is clear that the EMs need to communicate with each other, as well. This is done via a *domain manager*, which communicates through the INMS with the service and business systems. This hierarchical arrangement of the management system components is outlined in Figure 13.1, which generally follows the *Telecommunications Management Network* (TMN) terminology established by the ITU-T.[3]

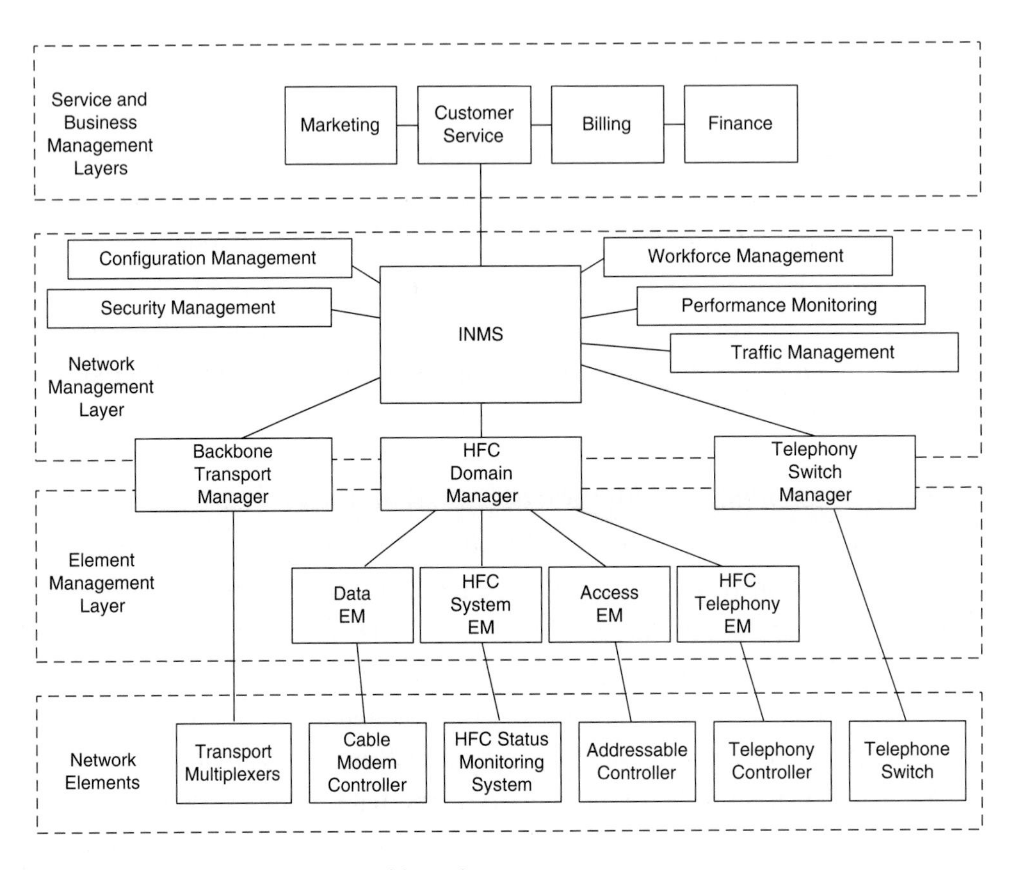

Figure 13.1 Network management hierarchy

At the top of the hierarchy are the *Service and Business Management Layers* (SML and BML), comprising the systems that connect with customers directly (customer service and billing) and that relate to the business operations (finance and marketing). These systems exchange information with the INMS, which is in many respects the heart of the overall system. The INMS resides at the *Network Management Layer* (NML), where the network security and configuration are controlled. Network performance and traffic intensity are monitored

3. V. Sahin, "Telecommunications Management Network, Principles, Models, and Applications," Chapter 3 in *Telecommunications Network Management into the 21st Century*, ed. S. Aidarous and T. Plevyak (New York, NY: IEEE Press, 1994).

at the NML. In addition, when a malfunction is detected, *trouble tickets* are generated at this level, work orders are issued, and service personnel are dispatched.

Not all functions need to be referred up to the level of the INMS, however. In fact it is good practice in network management to have analysis and control activities performed at the lowest practical level. Thus, for example, the HFC domain manager will perform fault isolation, where possible, and then simply report the results of its analysis up to the INMS, thus off-loading tasks from the INMS. Similarly, the domain manager enables communications between the services operating over the HFC plant and the plant itself, which don't need to go through the INMS.

Perhaps an even more important role for the HFC domain manager is to serve as an interim network management platform until the operator is ready to make the major investment for a full-blown INMS system. Domain managers are being developed that provide communications between plant and application EMs and that allow some connection with higher-level functions, such as maintenance dispatch systems.

We should take a moment to make clear the distinction between an HFC element manager and the traditional "status monitoring controller." The primary role of the status monitoring controller is to communicate with transponders and end-of-line devices and to carry out routine tests, such as FCC compliance monitoring. It also has a user interface for reporting and logging alarms. The HFC element manager uses those alarms to analyze the failure and to set remedial actions in motion. Typically the HFC EM display will present both the network connectivity (stick diagram of network topology) and the geographical locations of the elements (street map). This provides the basis, first, for fault correlation analysis and, second, for dispatch of maintenance forces. The HFC EM communicates with application EMs through the HFC domain manager.

Communication protocols and standards

The cable industry is moving as rapidly as possible toward open (non-proprietary) interfaces for all the products that it uses. There has been widespread agreement on the use of *Simple Network Management Protocol* (SNMP) for communications between the EMs and the HFC domain manager. SNMP is a very simple protocol with only four commands: SET (for the manager to set a parameter in an element), GET (for the manager to acquire data from an element), GET NEXT (for the manager to discover managed objects methodically), and TRAP (for the element to transmit an alarm, even when it has not been requested by the manager). In order to use SNMP, one must define documents called *Management Information Bases* (MIB). The MIB document has three main functions:

- To define the data interface for communicating with the network element
- To state exactly how data elements are filed, similar to the librarians' Dewey Decimal System
- To describe the information itself[4]

With the MIB for a system in-hand, anyone with the proper skills can write a program to monitor and manage the network element through its element manager. To permit the continued use of legacy status monitoring controllers not originally designed to communicate upward in SNMP, *SNMP proxy agents* can be developed that provide the appropriate interfaces. It has been more difficult to establish open and standardized protocols at the level between the NEs and the EMs, because significant numbers of legacy proprietary systems are already in place. In North America, CableLabs has been coordinating an effort to define open standards for both the communication protocols and the physical attributes of the transponders for HFC status monitoring. One of the objectives of the standardization is to reduce the cost of the transponders, which is deemed necessary to make system-wide deployments in amplifiers and nodes practical.

Figure 13.1 shows two other domain managers: for a telephony switch and for a backbone digital transport system that interconnects headends (Chapter Twelve). This reflects the reality that these types of products have well-established management systems deployed and functionally integrated into their operation. In general, because of protocol and operational differences, it is most

4. Sometimes people refer to the database itself as the MIB.

efficient to interconnect these management systems and the HFC domain at the INMS level, rather than to try a direct connection into the domain manager.

Figure 13.2 shows the SNMP "pipe" that connects the various EMs with the HFC domain manager. The physical connection is an Ethernet, which can also be used to connect each EM with its application controller. Proprietary communications between the NEs and EMs take place concurrently with SNMP communications between EMs and the domain manager.[5] Note that the HFC EM is shown merely as an SNMP proxy agent placed above the status monitoring controller. This reflects the method that is used for legacy (pre-SNMP) monitoring systems, with many of the EM functions actually being performed by the domain manager. This is generally the most efficient allocation of tasks for such cases, rather than developing two separate managers.

Distributed Element Management Systems

In the previous chapter we discussed that HFC network architectures are becoming more distributed. What may in the past have been five adjacent semi-autonomous operations, each with its own headend, is now a single system with one main headend feeding its distribution as well as four hubs situated where the other headends used to be. In this section we will discuss how the network management of these distributed systems is performed.

Since the system operator is likely to take both the plant integration and the network management implementation in steps, we will start at a relatively early stage of the evolution. Let's assume that there are only two sites and only status monitoring is implemented, but let's also assume that the HFC element manager is to be located at a third site (perhaps the business office). As shown in Figure 13.3, each of the operating headends has a PC connected via an RS-232 cable to a status monitoring controller.[6] Each PC is running the status monitoring program and has a remote call-in program, such as pcANYWHERE,[7] running, as well. With an ordinary telephone modem connection to the public telephone net-

5. Since SNMP was always intended to be used for critical management functions, it needs to be extremely robust. Accordingly, it supports only the UDP/IP protocol, which has very little overhead (see Appendix C). There is no problem with multiple protocols operating over a single Ethernet physical layer.

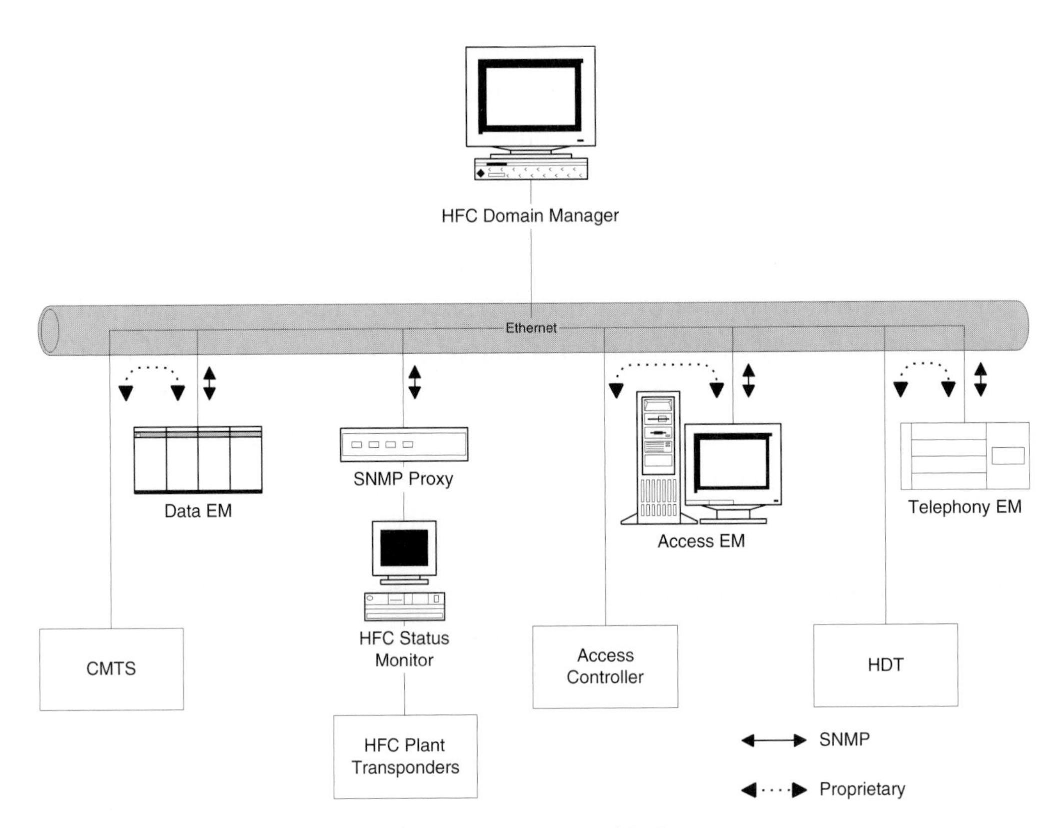

Figure 13.2 Interconnection of HFC domain manager with element managers

work, the operator then has the ability to dial in to see what is running on any of the local monitoring systems. The system can be configured to automatically dial up to the main computer to register alarms, or it can be equipped for beeper dial-up. This remoting scheme entails the cost of a telephone line.

In time, as more service applications are added at these headend sites, it will become more effective to connect all of the EMs in the site with an Ethernet and to bridge that LAN to a WAN that interconnects all the hubs to the main headend. The connections within each hub site will resemble that of Figure 13.2,

6. In a large system, multiple controllers will be combined through an RS-232 multiplexer to the PC. Also some monitoring systems use RS-485 for internal communications, which requires a 232/485 adapter, but that is a minor item.

7. PcANYWHERE (32 bit), Symantec Corp., Cupertino, CA.

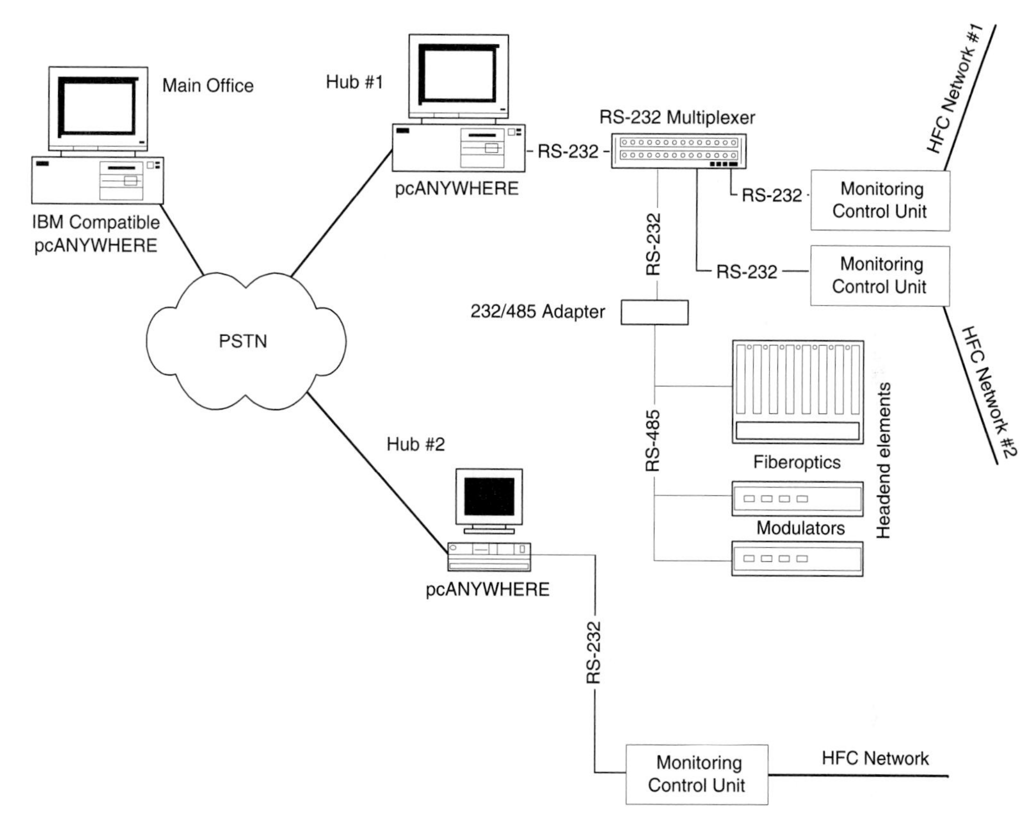

Figure 13.3 Remote monitoring

except that the domain manager will be located at the main headend. In the hubs, the Ethernet is merely bridged to the WAN. As noted in Chapter 12, as data or telephony services are deployed, the interconnect system will need to be quite robust in both directions. Hence it is likely to include SONET or SDH transport, which means that Ethernet ports on the WAN will be available at each site without necessitating separate leased telephone lines.

Return Path Performance Monitoring Systems

The performance of the forward path is usually monitored either with transponders in amplifiers and nodes or with *end-of-line monitors*. The transponders report whether the unit is functioning within pre-set limits, including an RF power measurement on one output port. Since the power of the forward signal should be well defined at all times, measuring power gives a good indication of C/N performance. End-of-line devices usually provide additional details on the forward spectrum,[8] but also rely on the well-established characteristics of the forward video signals.

Performance monitoring for the return path, on the other hand, is not as straightforward. Three types of information are available: transponder alarms indicating when either AC or DC power is out-of-spec, headend alarms for low received optical or RF power, and application receiver alarms for high error rates. Any of these will alert the operator to problems, but none of them is very helpful for troubleshooting the sorts of ingress and noise problems peculiar to the return path, or for heading them off before they threaten data transfers. When application receivers get unacceptable error rates, they merely command their in-home transmitters to send at higher powers. Typically, alarms are sent only when the high power limit is reached. By that point, the system could already be in deep trouble.

Since much of the return path traffic consists of bursty data, simple power measurements aren't meaningful indicators of in-service performance, so other methods must be found. The application receiver does acquire useful information, especially for communication systems with forward error correction (FEC, discussed in Chapter Three). That receiver could make available its information on changes in transmitter output power, the count of corrected errors, and changes in adaptive equalizer coefficients (Chapter Three). All of those data would be relevant to an evaluation of return network performance. Trends could be analyzed in a straightforward manner that would indicate incipient problems. If time stamps were recorded, as well, it would even be possible to triangulate the source of intermittent ingress by comparing several return paths.

It has been suggested[9] that specific messages could be transmitted from a set-top box or modem specifically for use in performance analysis. These mes-

8. Generally providing information sufficient for US operators to determine compliance with FCC performance criteria.

9. D. Pike, proposal to NCTA Engineering Subcommittee on Recommended Practices, May 1997.

sages would be preceded by a *unique word* (a particular data pattern not used in other communications) that would identify them as diagnostic tools. The communication system could be set up so that a downstream message would trigger the upstream message, which could then include both a downstream receive level (with C/N or error information) and an upstream transmit level.

Summing-Up...

- Increasing levels of complexity and of interconnection are making some degree of network management more and more of a requirement for successful operation.
- Considerable management can be accomplished at the domain manager level.
- Standardized communication protocols exist at the element manager level and above, but not at the level of the elements themselves.
- HFC network performance monitoring is likely to be done most effectively by collecting and coordinating communication performance information from the applications running over the network.

APPENDIX A

Measuring Digital Signals with a Spectrum Analyzer

This appendix is intended to help cable engineers and technicians who are familiar with standard analog measurements on a spectrum analyzer to bridge over to the digital world. Our intent is to steer you around some of the measurement "potholes" we have already found. Fuller discussions of these items should be available from instrumentation manufacturers. There is also a comprehensive text on this subject.[1]

True Average Power Meter

The most accurate method for measuring the power of a digital channel is to put the signal into a true average power meter. That meter actually uses a thermocouple to measure the heat created by the signal. This means that it can provide the correct average power value, regardless of bandwidth or peak factor, provided that the signal of interest is the only signal present at the meter input. In the field, however, it is rare to have access to individual signals, and true average power meters are not usually built for use outside of labs, so this is not a tool likely to gain widespread acceptance in the field. Keep in mind, however, that

1. J.L. Thomas, *Cable Television Proof of Performance* (Englewood Cliffs, NJ: Prentice Hall PTR, Inc., 1995).

the true average power meter is the most accurate instrument, and it can be used to determine the accuracy of any other method.

Spectrum Analyzer

Spectrum analyzers are designed to measure the average power of sinewaves. Accordingly, there will be some inaccuracies associated with measurements of any other type of signal. In this section we will cover several common methods of measuring digital signal levels—normal marker, noise marker, and channel power measurement algorithm—and will discuss the pros and cons of each.

Spectrum analyzer marker

Our first temptation in measuring a signal with a spectrum analyzer is to put a marker on the signal and read the value. An example of this is shown in Figure A.1. Unfortunately, the reading given by the marker is incorrect. This can be seen by comparing Figure A.1 with Figure A.2. The only change in Figure A.2 is to use a resolution bandwidth of 30 kHz instead of 10 kHz. Notice in the second figure that the level of the signal (as reported by the marker) has changed from 8.2 to 13.1 dBmV, even though the signal itself did not change. Also notice that the entire waveform has moved up on the display (but neither the signal nor the reference level had actually been changed). The reason the signal levels appear to have changed is that the spectrum analyzer measures total composite power of all signals that fall inside the resolution bandwidth filter. As the resolution bandwidth filter is increased (from 10 kHz to 30 kHz in this example), the amount of energy within the filter bandwidth increases, which, in turn, causes the displayed level to increase.

Those familiar with measuring C/N with a spectrum analyzer will probably recognize this phenomena and realize that it is possible to correct for the error caused by the resolution bandwidth. We want the measurement to show the total power of the digital channel. Therefore, we need to convert from the displayed power to channel power. The equation is

$$\text{Channel power} = \text{Displayed power} + 10 * \log\left(\frac{\text{Channel BW}}{\text{Resoln BW}}\right) + \text{Correction}.$$

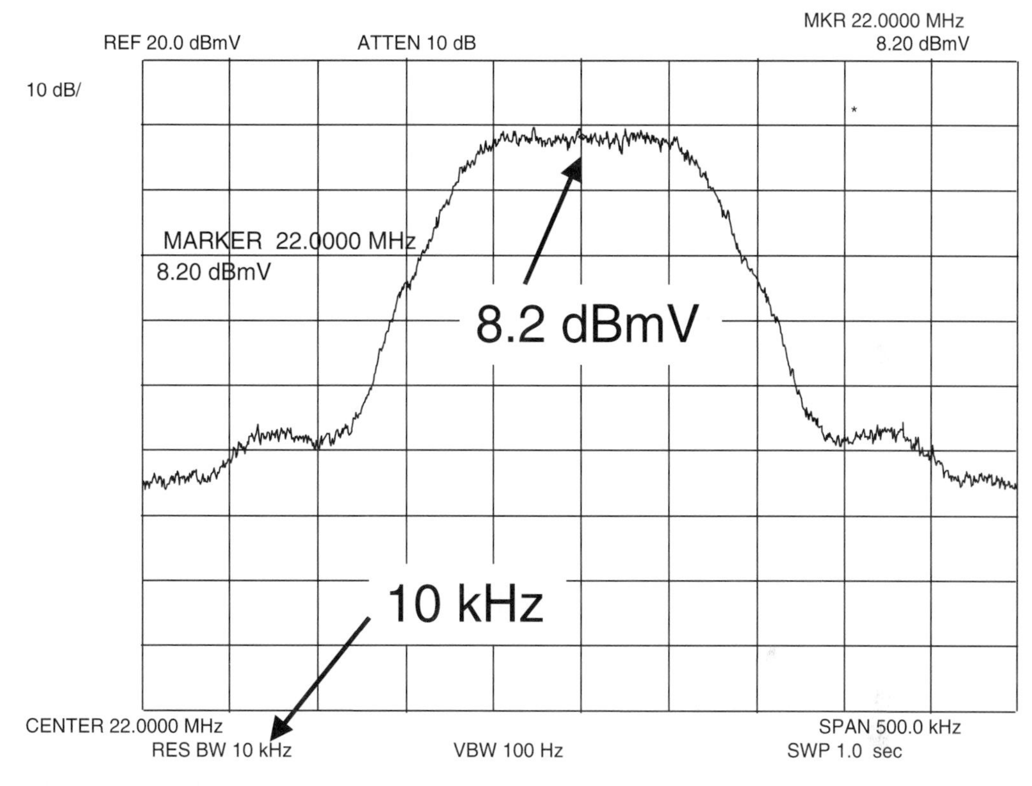

Figure A-1 Measurement of digital signal

The correction factor is required because the amount of noise captured by the resolution bandwidth filter is not the same as for the filter's specified 3 dB bandwidth. In other words, the noise bandwidth of a 10 kHz filter is not actually 10 kHz. Another error included in the correction factor accounts for the log amp that converts the input to a logarithmic display. It does not properly convert noiselike signals. The total correction factor is usually around 1 to 2 dB and can be obtained from the spectrum analyzer manufacturer.

When performing a measurement with a spectrum analyzer, the sample detector should be used, rather than the peak detector, and the video bandwidth should be set low enough that the signal appears fairly smooth and the measurement does not vary significantly from sweep to sweep.

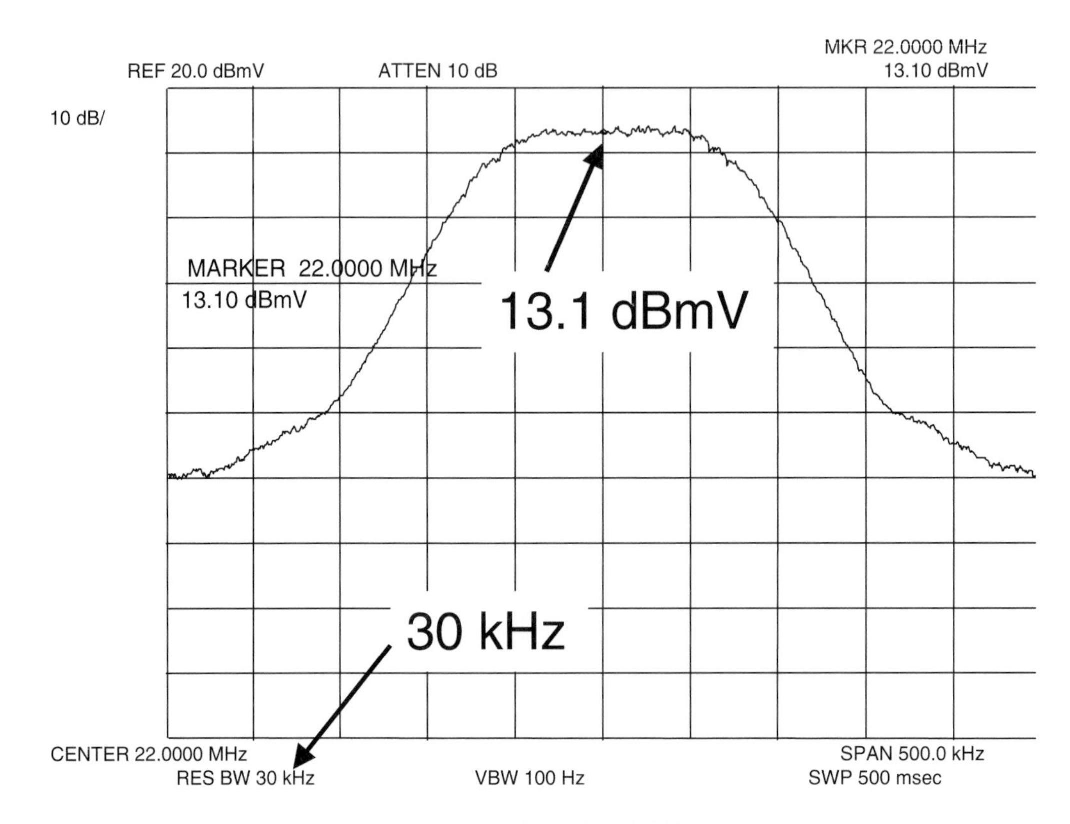

Figure A-2 Measurement with increased resolution bandwidth

Spectrum analyzer noise marker

A better way to measure noiselike signals, such as the digital signals we have been describing, is to use the *noise marker*, a feature that is included on most newer spectrum analyzers. It compensates for the errors described in the previous section by displaying power in a 1 Hz bandwidth, as shown in Figure A.3. This allows the channel power to be calculated by the following equation:

$$\text{Channel power} = \text{Displayed power} + 10 * \log(\text{Channel bandwidth})$$

No other correction factors are required. In addition, the resolution bandwidth setting does not change the value displayed by the noise marker.

One precaution must be observed when using the noise marker with certain spectrum analyzers.[2] The signal being measured must fill several graticles (major divisions) of the display. This is because the value displayed by the noise marker is

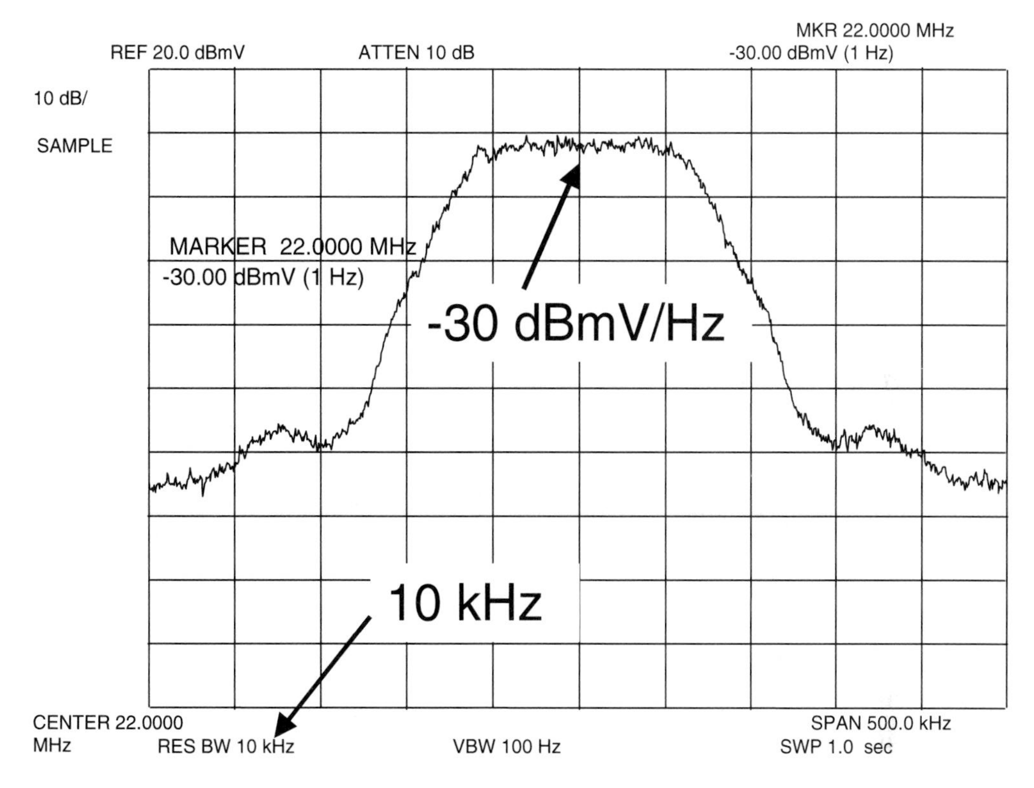

Figure A-3 Noise marker measurement

actually the average value of all the noise within about 1 to 2 graticles of the marker. Figure A.4 shows a signal that is being improperly measured. The actual signal is the same one shown previously in Figure A.3, but the span has been increased so that the signal now appears very narrow and does not fill even one entire graticle. Observe that the noise marker thinks the power is –32.2 dBmV/Hz instead of –30 dBmV/Hz. This is because the noise marker is averaging in some of the spectrum next to the channel, which does not contain any signal power. If you look carefully at the figure, you can see that the marker actually appears slightly below the signal.

2. This caution applies, for instance, to Hewlett-Packard analyzers. Products from some other vendors, such as Tektronix, make the required adjustments automatically.

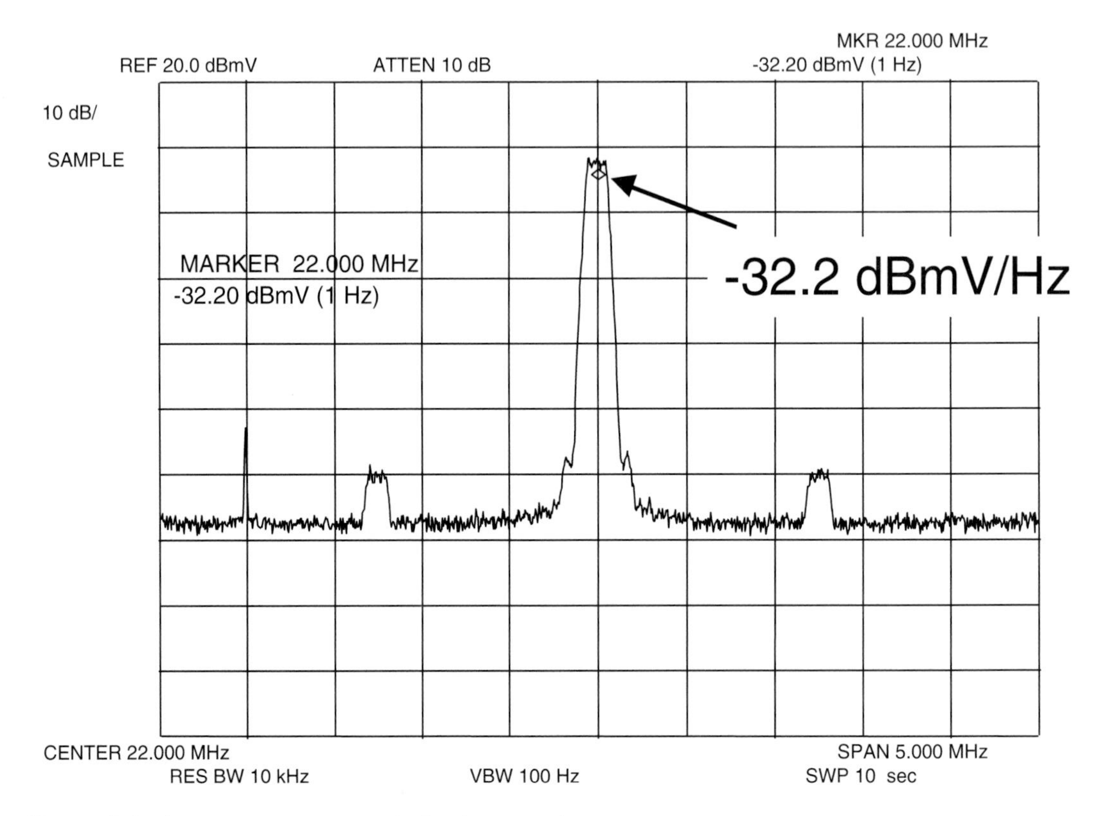

Figure A-4 Improper measurement, due to excessive span

Channel power measurement algorithm

Some of the newest spectrum analyzers have algorithms specifically designed to correct for all the errors mentioned previously and display the actual power of the digital channel directly. This feature can be useful in avoiding misleading measurements. Follow the manufacturer's suggested procedure for making the measurement.

Summary

The true average power meter is the only instrument that will give the absolute correct answer for the power in a digital signal. When applied correctly, a spectrum analyzer with channel power measurement feature or noise marker can also give very accurate results. Perhaps the best and simplest way to make sure that everyone gets the same answer when measuring a signal is to compare the result of your favorite method to the reading from an average power meter. The difference between the two methods could become a system-wide correction factor. As long as the type of signal being measured and the equipment being used (along with the settings used on that equipment) do not change, the correction factor will be valid.

APPENDIX B

Details of Design and Setup

B1 – Unity Gain and the Unity Gain Point

Unity gain in a cable TV system means that the gain of each amplifier station exactly duplicates the total loss between and inside the stations. Since the loss equals the gain, the net gain from station to station is 0 dB. The term "unity" refers to the number one. A "gain of 1" is another way of saying a "gain of 0 dB." Thus, the term "unity gain" refers to a gain of 0 dB. Figure B1.1 shows how this applies in the forward path.

Figure B1.1 shows forward path signal levels. At each amplifier station, the output level is +47 dBmV at the highest channel. All stations are set to the same level. The gain from one station's output to the next station's output is 0 dB or a "gain of 1," thus "unity" gain. During forward path setup, the gain of each station is adjusted so it exactly matches the loss of the cable and passives that preceded it.

Why is it important to have unity gain? In the forward path, we can see that unity gain assures that the same level comes out of each amplifier. When the system is designed, the tap values are chosen with the assumption that every amplifier will have the same output level. Since the tap values are chosen to provide a consistent loss (at the highest channel) between the output of each amplifier and each tap port, it is important that the return path have that same relationship between a level at a tap port and the level at the amplifier.

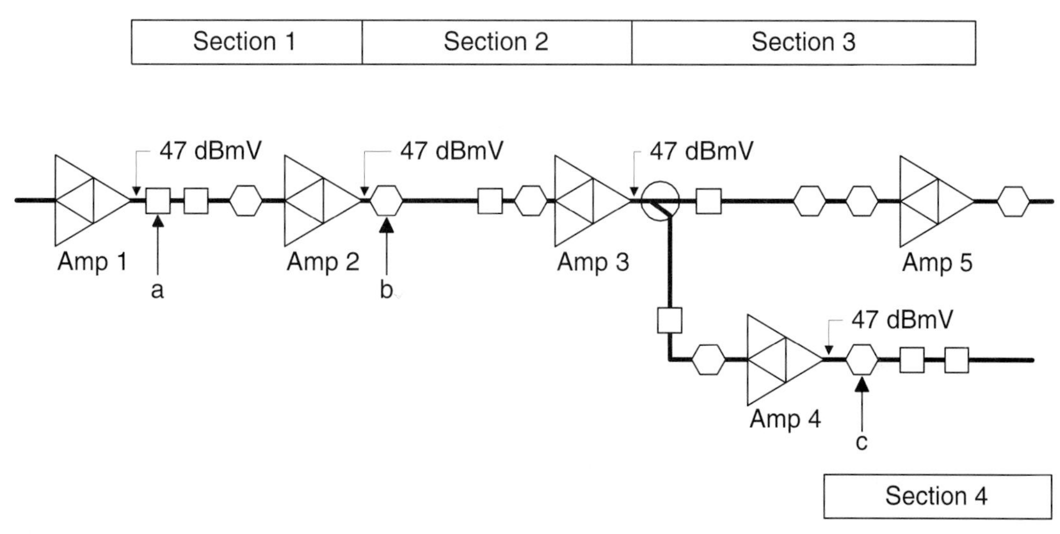

Figure B1.1 Unity Gain Example

Perhaps an example will clarify the situation. Let's assume that amplifiers 1, 2, and 4 in Figure B1.1 are followed immediately by a 26 dB tap. Let's further assume that the homes connected to these 26 dB taps have a set-top box that is transmitting at the proper level to provide 36 dBmV at the tap port. As a result, each set-top box will provide 10 dBmV at the amplifier station. Let's further assume that 10 dBmV at the amplifier station provides 10 dBmV at the headend. Now, if the plant is not set to unity gain some strange things will happen. Let's assume that Section 2, which includes the return path circuit in Amp 3, has 5 dB too little gain. The signals injected at points "a" and "b" will still arrive at the headend at 10 dBmV. However, the signal injected into point "c" will go through 5 dB too little gain on its way to Amp 2. Therefore, it will arrive at Amp 2 at 5 dBmV and will, therefore, arrive at the headend at 5 dBmV.

When a non-unity gain situation exists in the return path, several things could happen.

- If the set-top boxes (or whatever else is using the system) are not adjustable, their signals will arrive at the laser at the wrong level. As a result, distortion or poor carrier to noise could occur.

- If the set-top boxes that transmit through the unity gain error are adjustable but are out of adjustment range, the same problem will occur.
- If a designer wanted to allow for such a gain error, he would need to back off all the other levels in the system by the amount of the error. This would make the levels in the entire plant operate closer to noise and ingress.
- If the set-top boxes are adjustable and still have a sufficient adjustment range, there will not be any problems.

Location of unity gain point

We have established why we need unity gain between amplifier stations, but we have not yet determined where the unity gain reference point is within the amplifier station. At first this might not seem to be a big issue, but if the correct location is not chosen and the alignment is not performed with the proper unity gain point in mind, we will experience the same types of problems that occur when the plant is not properly adjusted for unity gain. We will show in this section why the best unity gain point is at the station's output diplex filter (return path input diplex filter).

To illustrate the problem, let's look at the block diagrams of two very different types of amplifiers, a line extender, shown in Figure B1.2, and a distribution amp, shown in Figure B1.3. For the sake of clarity, the diplex filters are assumed to have zero loss in these examples.

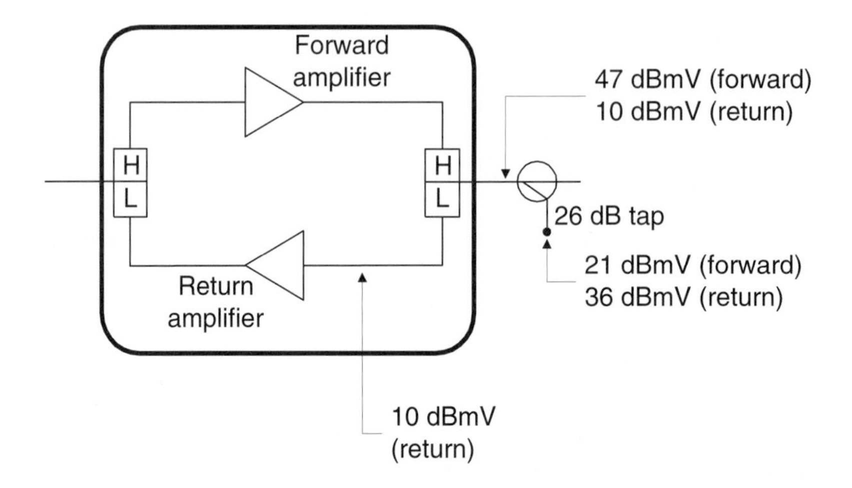

Figure B1.2 Line extender

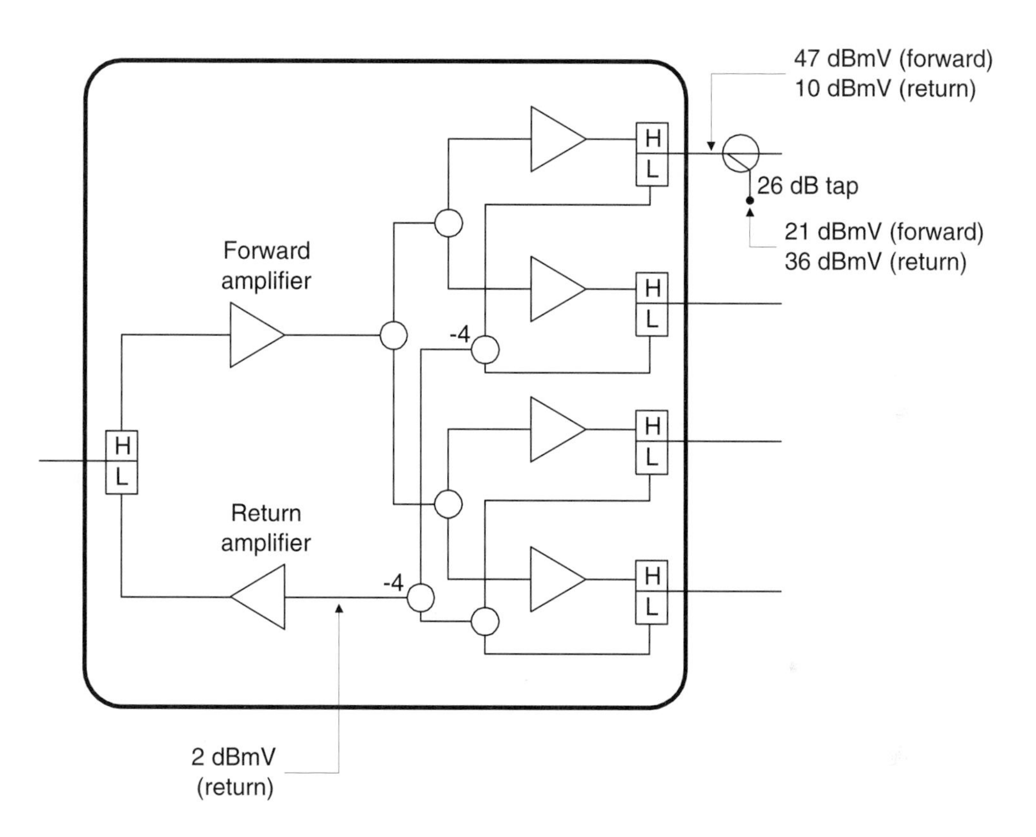

Figure B1.3 Distribution amplifier

Both Figure B1.2 and Figure B1.3 show a signal being injected into a 26 dB tap that is connected to the station's output port. In both cases the injected signal is 36 dBmV at the tap port. After passing the 26 dB tap loss, both signals arrive at the diplex filter at 10 dBmV. Since a constant level is required at every diplex filter, the plant is properly aligned. In other words, 36 dBmV is required at a 26 dB tap, regardless of the amplifier type.

An alternative would have been to define the unity gain point to be at the input to the return hybrid instead of at the diplex filter. If the level at the hybrid input is a constant (10 dBmV, for instance) for all types of amplifiers, however, application transmitters in subscribers' homes would have to operate at different levels, depending on what type of amplifier they were feeding into. There would be no input change for a line extender since there is negligible loss between the diplex filter and the hybrid input, but a distribution amplifier would operate at very different levels. Figure B1.4 shows a distribution amplifier with these levels.

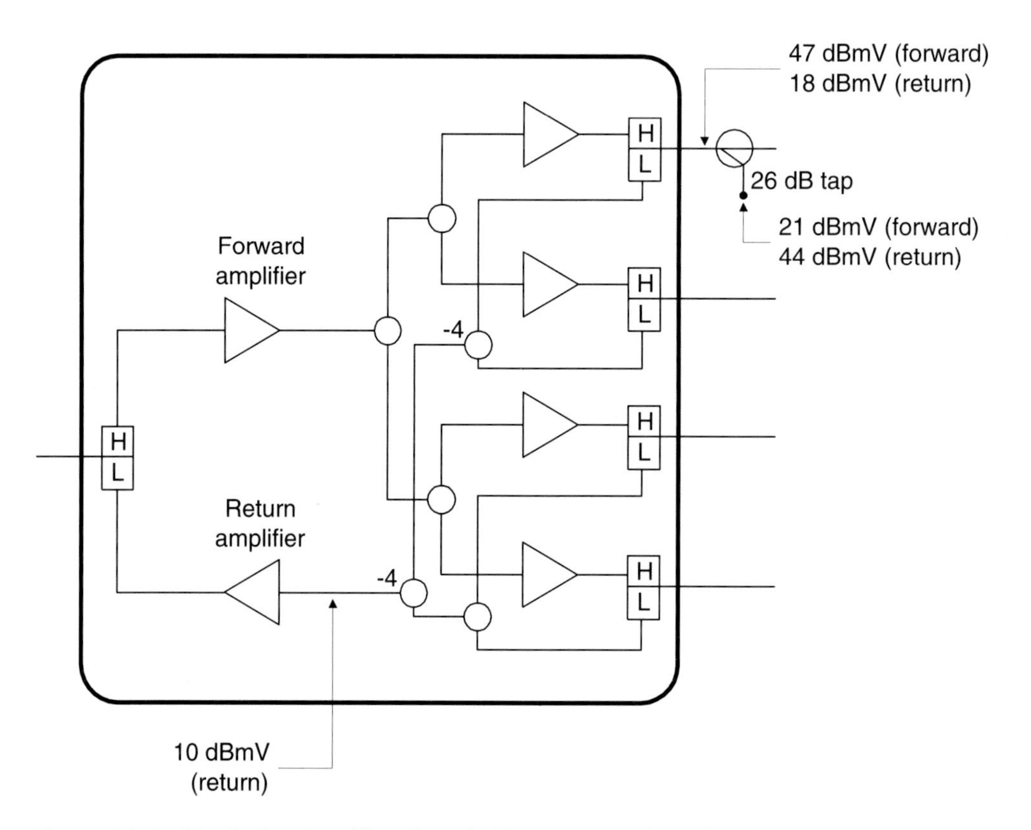

Figure B1.4 Distribution Amplifier aligned with incorrect unity gain point

Notice that now the terminal equipment needs to transmit at an 8 dB higher level in order for the hybrid input to be 10 dBmV. As a result, the terminal equipment that feeds a distribution amp would need to be operated higher than the same equipment feeding a line extender. But since 50 percent or more of the homes feed into line extenders, this will tend to depress the operating levels of the majority of in-home transmitters. As is discussed in Chapter Eleven, the carrier-to-ingress of the system is highest when the in-home transmitters are operating as high as possible.

Furthermore, if the diplex filter is used as the unity gain point, then the amount of return gain for all line extenders in the system will be the same, independent of whether they are feeding another LE or a 4-port distribution amplifier. Using the hybrid input as a reference point would require different gain modules, depending on what type of amplifier is being fed upstream. This would be a logistical headache that would make pre-configuring amplifiers for two-way operation almost unmanageable.

Throughout this discussion, we have been referring to the best unity gain location as being at the diplex filter. In general, this statement can be simplified, and one can say that the unity gain point is the station return input port (same as the station forward output port). However, there are station topologies that allow for plug-in options between the diplex filter and the station port. Since such an option will "behave" like feeder loss, it needs to be treated like feeder loss. This is illustrated in Figure B1.5.

The plug-in option in Figure B1.5 is a 9 dB directional coupler. The coupler is assumed to have a 1 dB loss in the through leg and a 9 dB loss in the coupled leg. The figure shows that the tap value selected for the feeder will be different because of the coupler (a 25 dB tap is shown in order to make the math simple, even though a 25 tap might not exist). The different tap value is selected so that the forward level at the tap port is constant (21 dBmV in the example). Similarly, a 17 dB tap is selected for the feeder fed by the tapped leg of the directional coupler in order to maintain a constant forward level at its tap port. Figure B1.5 also shows a constant return path injection level of 36 dBmV into each tap port. Remember that the aim of our setup is for each house to have the same nominal design level for return path injection into the tap port. As a result of the directional coupler, the return path levels at the station are different depending on which leg of the coupler is being measured. Nevertheless, the level at the diplex filter is still the same 10 dBmV that we have been working with all along.

This example shows why the best definition of unity gain point is at the station output diplex filter. In most cases there are no significant losses between the filter and the station port, which means the unity gain point is effectively the station port.

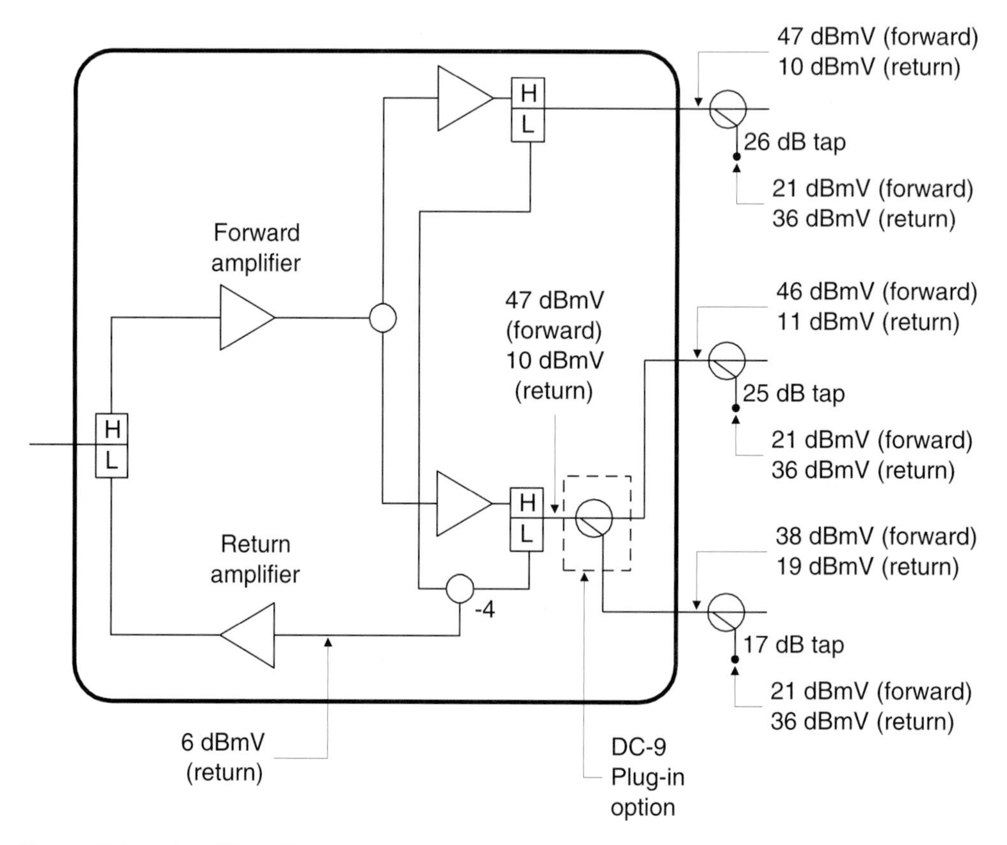

Figure B1.5 Amplifier with plug-in option at output

B2 – Sections Where the Forward Levels Are Not Standard

There may be situations where one section of the plant needs to be operated at different forward levels from the rest of the plant. Let's consider a case where the forward signals are lower. This will mean that the tap values will also be lower and, as a result, the return input levels will be too high unless we make some correction. If no special care is taken, the transmitters at the home will be instructed to turn down their levels. This is undesirable because the transmitter levels will be getting closer to ingress. Alternatively, if the forward levels in a section of plant are higher than the rest of the plant, the tap values will be higher and the transmitters will get instructed to adjust to higher output levels that some transmitters will not be able to reach. To solve this problem, whenever forward output levels are different, the return path input level should be changed by the same amount in the opposite direction. Let's illustrate with the line extender. The original levels are shown in Figure B2.1.

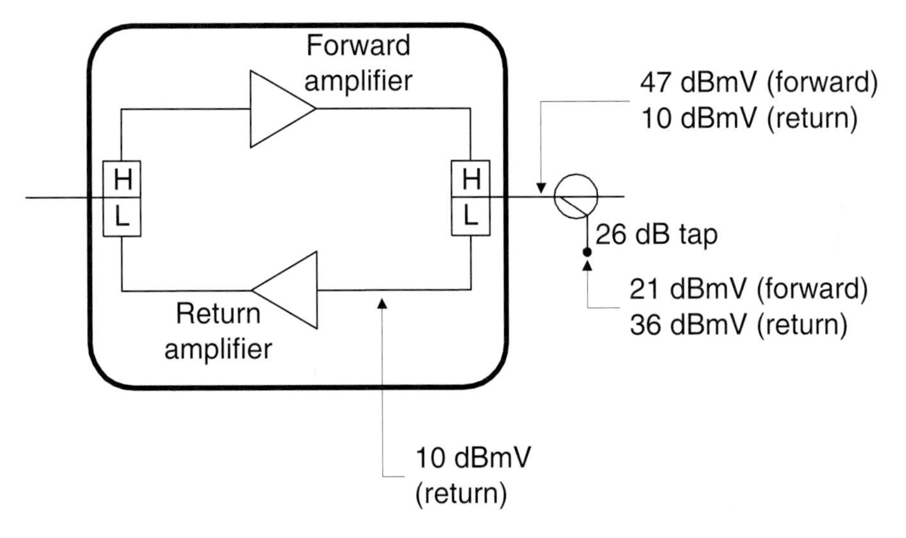

Figure B2.1 Original levels

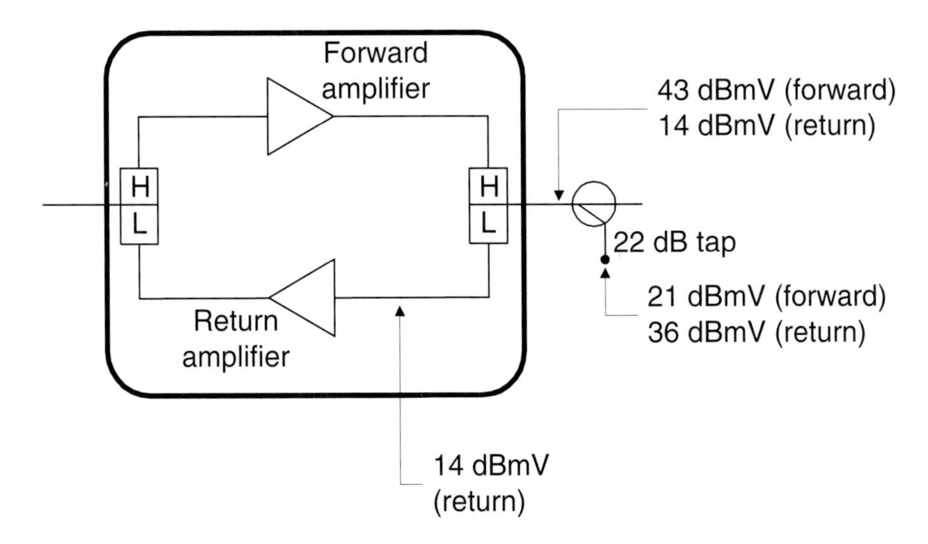

Figure B2.2 Forward path levels lowered 4 dB

Figure B2.2 shows what will happen if the forward path levels were lowered to 43 dBmV for a section of the plant. The figure shows that a lower value tap would be selected because of the lower forward path output level. When a constant tap port injection level is maintained, the level will be 4 dB too high at the diplex filter. To compensate, the return path level should be aligned 4 dB higher at this station. There are several ways to accomplish this:

- Manually add 4 dB to the injection signal.
- Look for 4 dB less gain on the sweep.
- Set the injection loss setting 4 dB lower (if the sweep setup has an injection loss setting).
- Subtract 4 dB from the "insertion point loss" given in the injection table that has been provided (such as Table 11-7 discussed in Chapter Eleven).

In summary:

If the forward path levels go <u>down</u> by X dB, then you can:

1. Raise the injection signal by X dB.
2. Look for X dB less gain on the sweep.
3. Set the injection loss setting on the sweep system X dB lower.
4. Subtract X dB from the "insertion point loss."

If the forward path levels go <u>up</u> by X dB, then you can:

1. Lower the injection signal by X dB.
2. Look for X dB more gain on the sweep.
3. Set the injection loss setting on the sweep system X dB higher.
4. Add X dB to the "insertion point loss."

B3 – Return Levels in Trunk Amplifiers and Express Feeders

One more unique situation that needs to be discussed is a trunk amplifier. Trunk amplifiers (and trunk node stations) have separate trunk and bridger output modules. The return paths of these modules are combined either through a combiner or a directional coupler. A simplified block diagram of a trunk station return path is shown in Figure B3.1.

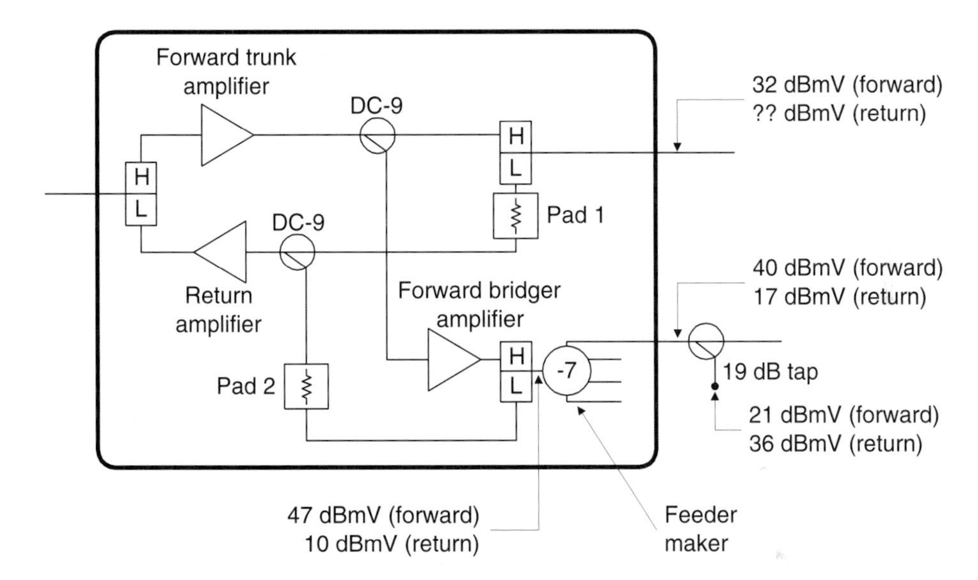

Figure B3.1 Trunk amplifier

Notice that in the feeder system (bridger output ports) the same levels still hold true. We still have 10 dBmV at the unity gain point which is the diplex filter. If pad 2 is zero dB, then we will have 3 dBmV at the return path amplifier after going through the 7 dB directional coupler. The trunk output, however, causes some grief. If we were to follow the rules given previously for differing port output levels, then we would expect 15 dB more signal at the trunk diplex filter than at the bridger diplex filter (since the forward path levels are 15 dB lower). That would give 25 dBmV at the diplex filter, which would require that Pad 1 be a 22 dB pad in order to get 3 dBmV at the hybrid input. Remember, it is necessary to have all ports hit the hybrid at the same level because—once the signals are combined—they cannot be adjusted independently. Since a 22 dB pad is unrealistic, there are several options:

1. Use a device called a bridger gate switch in the bridger return path (near the Pad 2 location). The bridger gate switch has a 7 dB gain. This will raise the level at the hybrid to 10 dBmV and will require only a 15 dB pad for Pad 1.

Note that this is not quite the same device as the ingress control switch (ICS) discussed in Chapter Five, although the terms are often used as though they are interchangeable.

2. Since there are usually no taps in a trunk line, the trunk return can be run at a lower level. For instance, the trunk could be run at 10 dBmV in spite of the fact that the forward path level is no longer 47 dBmV. In this case, pad 1 could be set to 7 dB, and the levels at the trunk diplex filter would be aligned just like the levels at the bridger diplex filter. The key point here is that the ratio of forward to return levels in the cable between stations is important only when there are taps in the cable. This is because our objective in setting RF plant levels is to optimize the output levels of in-home transmitters. No taps means no transmitters, hence no concern.

Express (untapped) feeders give rise to another situation where forward path levels are different for a section of the plant. In the example shown in Figure B3.2, the forward levels on the express feeder are set at 44 dBmV, while the tapped feeders are driven at 51 dBmV. We could apply the rules from Appendix B2 and align the return path at different levels through the express feeder, but this turns out to be an unnecessary complication. As we have just pointed out for the trunk amplifiers, since there are no taps in the express section, we do not have to be concerned about maintaining a specific ratio between forward and return levels. It will be fine to simply align the return to the same port levels as for the tapped feeder.

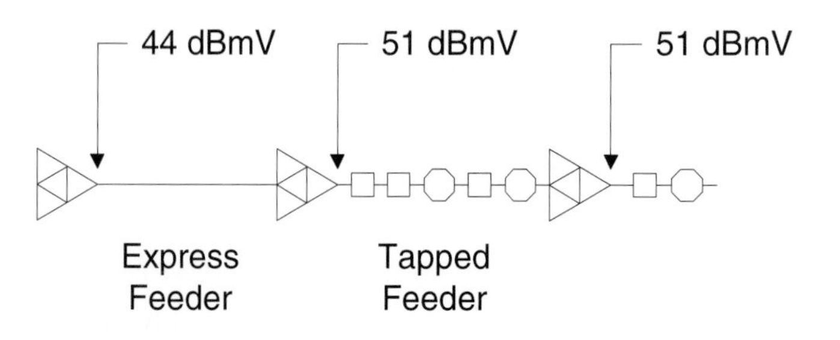

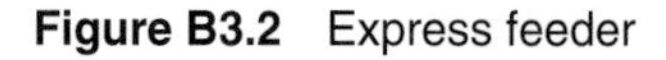
Figure B3.2 Express feeder

B4 – Transporting Analog Video Upstream

There are situations where an operator needs to bring video signals back to the headend in analog form. If the return path is expected to be used for interactive applications, it is unlikely that it will be cost-effective to allocate the requisite laser power to a single video service. On the other hand, it is possible to carry good-quality analog video on return lasers if the video service is either not sharing the laser with any other services or if the other services can be limited in such a way that they do not interfere. We can state this as a rule:

> *Whenever an analog video channel is present, the payload must be specified such that no deleterious distortion products fall within the video band. As long as this requirement is followed, any combination of NTSC video and digital data may be used.*

Specifically, this means that if lower-cost uncooled lasers are used, one must select the video channel locations carefully so that distortion products related to other return services do not fall within the video band.[1] When analog and digital data are carried on the same laser, the recommended total power still needs to be observed. We should take a moment, first, to clarify some video terms and to connect them to the language used elsewhere in this text. We use "total power input" to specify the total average power presented to the input of a laser. The term "analog video input level" is a term based on the "sync tip level." Technically, the sync tip level is the average power (RMS voltage) of a sinewave that has the same amplitude as the video carrier during the tip of the sync signal. Although this may sound complicated, it is the way video signals have been measured for 40 years. Since analog video consists of downward modulation, the average power of a modulated video signal is lower than the average power of the carrier. As a result, the average power of a modulated video carrier varies from about 3 dB to 8 dB less than that of an unmodulated carrier, depending on video content. It is conservative to assume that the average

1. Video performance is usually limited by second-order beats, since the third-order distortions from return lasers are lower.

power of a modulated video signal is 4 dB less than the unmodulated carrier. Therefore, if a video signal is inserted at a video input level numerically equal to the recommended total power input level for the laser, its average power is at least 4 dB below. When two video channels are inserted, the total power is still 1 dB below the recommended total power input level.

At such input levels, there may still be some power available for light data loading, providing that the previous rule on second-order beats can be adhered to. Let's assume that 45 dBmV is the total power rating for the laser, and that two video signals are applied at video input levels of 45 dBmV each. Then

		Explanation
Recommended Total Input Level	+45	Laser specification
Sync Tip Level of Each Video Channel	+45	Input specification
Average Power of Each Video Channel	+41	Avg power 4 dB lower than sync tip
Total Avg Power of Two Video Channels	+44	Power addition: 10*log2

By referring to the power addition table (Table D-2 in Appendix D) we can determine that +44 dBmV added to +38 dBmV will combine to +45 dBmV. This means that even with two videos present, a data payload with a total power of 38 dBmV can be added.

In order to understand what this means, recall that for similar lasers we have allocated 45 dBmV total power to 35 MHz of data on a constant power per Hz basis. This is a power density of $45 - 10*\log(35 \times 10^6) = 45 - 75.4 = -30.4$ dBmV/Hz, or +29.6 dBmV/MHz. We can determine how much bandwidth would be associated with 38 dBmV by solving thus:

$$29.6 \text{ dBmV/MHz} + 10*\log(X) \text{ MHz} = 38 \text{ dBmV}$$
$$\log(X) = (38 - 29.6) \div 10 = 0.84$$
$$X = 7.0 \text{ MHz.}$$

Therefore, there is enough power available for 7.0 MHz of data, along with the two video channels. Remember, however, that all data and video must be located in areas of the spectrum that do not produce any second-order beats inside the video band.

Baseband Digital Systems

Only people from the cable broadband industry need to add the word *baseband* before the word "digital" when referring to digital signals that are not being transported on modulated carriers. The rest of the world is often completely unaware that non-baseband (modulated carrier) digital signals even exist, even though that is how they are communicating every time they use a telephony modem. Throughout this book, when we have discussed transport of digital signals, our subject has been modulated carrier digital. This appendix will deal exclusively with baseband digital transport, however. In this brief overview, our aim is limited to giving the reader:

- a starting level of understanding of some of the key transport systems that are being employed along with HFC networks, for those readers who have not yet been exposed to these technologies, and
- a convenient reference for some of the important quantities in the standards for these systems.

C1 – Backbone Transport (SONET and SDH)

The first of our subjects is the two high-speed transport standards, SONET and SDH, that are used in headend interconnects and in public telephony transport networks. SONET is an acronym for Synchronous Optical Network and was developed by North American telephony operators and suppliers.[1] SONET-compliant equipment is being applied throughout Canada, South Korea, Taiwan, and the United States. A variant, SONET-J, is used in Japan. SDH, which stands for Synchronous Digital Hierarchy, was developed within the ITU[2] and is the standard for transport essentially everywhere else in the world.

Both systems specify rates and other parameters for time division byte-interleaved multiplexes of information streams. Two important aspects of these systems make them far superior to all of the telephony transport hierarchies that preceded them. The first is that the systems were designed to facilitate switching. This was done by allowing any particular lower subunit of a high-level time multiplex to be extracted and replaced without disassembling the whole stream. This is possible because all of the data rates in these systems are exact multiples of a base rate and because synchronization information flows through the networks along with the payload (as will be discussed later).

As shown in Table C1-1, a 2.5 gigabit per second (OC-48) SONET multiplex contains forty-eight 51.84 megabit per second (STS-1) subunits,[3] called *tributaries*. If at some network node there is a need to drop one of the STS-1s and replace it with a different one, a SONET *add/drop multiplexer* (ADM) can pick out that one 51.84 Mbps stream and insert the new one without touching the other 47 streams within the 2.5 Gbps multiplex. In addition, SONET pro-

1. Synchronous Optical Network (SONET) Transport Systems: Common Generic Criteria, TR-NWT-000253, Issue 2, Bellcore, Red Bank, NJ (December 1991).

2. "Characteristics of Synchronous Digital Hierarchy Multiplexing Equipment Functional Blocks" (ITU-T G.783) and "Network Architecture" (ITU-T G.803 and G.831), International Telecommunications Union, Geneva. http://www.itu.ch

3. STS stands for Synchronous Transport Signal. Note that the STS-1 rate is somewhat higher than the previous US standard rate of DS3 (44.7 Mbps). Thus, simple interfaces have been developed to permit continued use of legacy DS3 equipment, with only a small loss of payload efficiency. SONET path, section, and line overhead bytes are added to the DS3 signal to fill it out to the full STS-1 tributary rate. Similarly, the 144 Mbps E-4 rate used outside North America fits inside the SDH STM-1 rate (Table C1-2). Note that the SDH hierarchy starts at 155 Mbps.

vides for substructures, called *virtual tributaries* (VT), that allow, for instance, 1.7 Mbps streams containing individual DS1s to be separated out of an STS-1 data stream without demultiplexing, just as the STS-1 can be extracted from an OC-n without demultiplexing.

As its name implies, the hierarchy relies on fiberoptics, without which the Gbps rates would be unthinkable (for unrepeatered distances exceeding 100 km). The STS-1 designation refers to the electrical data signal at 51.84 Mbps; when converted to its optical form for transport over fiber, it is referred to as an OC-1.

Table C1-1 SONET/SDH hierarchy

Bit rate	Optical stream	SONET electrical	SDH electrical
51.84 Mbps	OC-1	STS-1	
155.52 Mbps	OC-3	STS-3	STM-1
622.08 Mbps	OC-12	STS-12	STM-4
1.244 Gbps	OC-24	STS-24	
2.488 Gbps	OC-48	STS-48	STM-16
9.953 Gbps	OC-192	STS-192	STM-64

Note: The OC-n bit rates are exact multiples. The values in the table have been rounded.

The second important advance is that the development of the two international standards was done cooperatively, although separately. Thus the rates, data characteristics, and signaling are mutually consistent. This means that SDH streams can be transported to (or through) North America without disassembly and reassembly at the "port of entry" and vice versa for SONET streams above the STS-1 rate. As can be seen from Table C1-1, the rates are consistent. The significance of the payload is specific to the system of origin only below the OC-3 rate.

These systems typically provide high degrees of reliability as a result of both component and route redundancy that is inherent in the equipment and the architecture. All transmissions proceed simultaneously "east" and "west." All electronics are backed up with a working spare, in a one-for-N arrangement with N = 4 to 12.

For video transport, the synchronous networks assure that a video payload will not get disassembled and recombined in an unpredictable fashion, as could be a concern with non-synchronous transport over the public switched network.

SONET and SDH represent the first truly non-proprietary telephony systems. This allows multiple vendors to develop whole systems or subsystems that best use their hardware/software design and manufacturing abilities. "Mid-span meets" have been demonstrated, proving interoperation between vendors. Since the network monitoring is also standardized, a transmission system can be used simultaneously for multiple services and customers, but with each customer able to monitor its own service independently.

SONET transport is structured in *frames* of bytes (8-bit words) at the STS-1 rate (Figure C1.1). The frames can be multiplexed with other data into higher rates or *concatenated* (strung together) for continuous data.[4] A SONET frame consists of 90 columns of 9 bytes each. The first three columns, referred to as the section and line overhead, are used for framing, routing, synchronization, and management. The remaining 87 columns are called the *synchronous payload envelope* (SPE). The first column of the SPE is path overhead, so 86 columns (774 bytes) remain for actual data payload. Since the STS-1 frame period is 125 μsec/frame, the payload data rate is 49.536 Mbps. Virtual tributaries are constructed out of full columns in the frame. The VT that carries a DS1, called a VT-1.5, uses three columns, for example.

It is relatively easy to see how a confined system could be run synchronously using a master clock, but it is harder to see how a dispersed system can avoid intersystem "boundaries" that would introduce slips in synchronization.[5] From the time these systems were first conceived, it was out of the question to

4. Concatenated STS-1s and OC-3s are denoted STS-1c and OC-3c, respectively.

5. The SONET and SDH standards were established before the satellite-based *Global Positioning System* (GPS) was ubiquitous, so GPS clocks were not an option. At the present time, however, some carriers have replaced their rubidium and cesium master clocks with GPS-based systems.

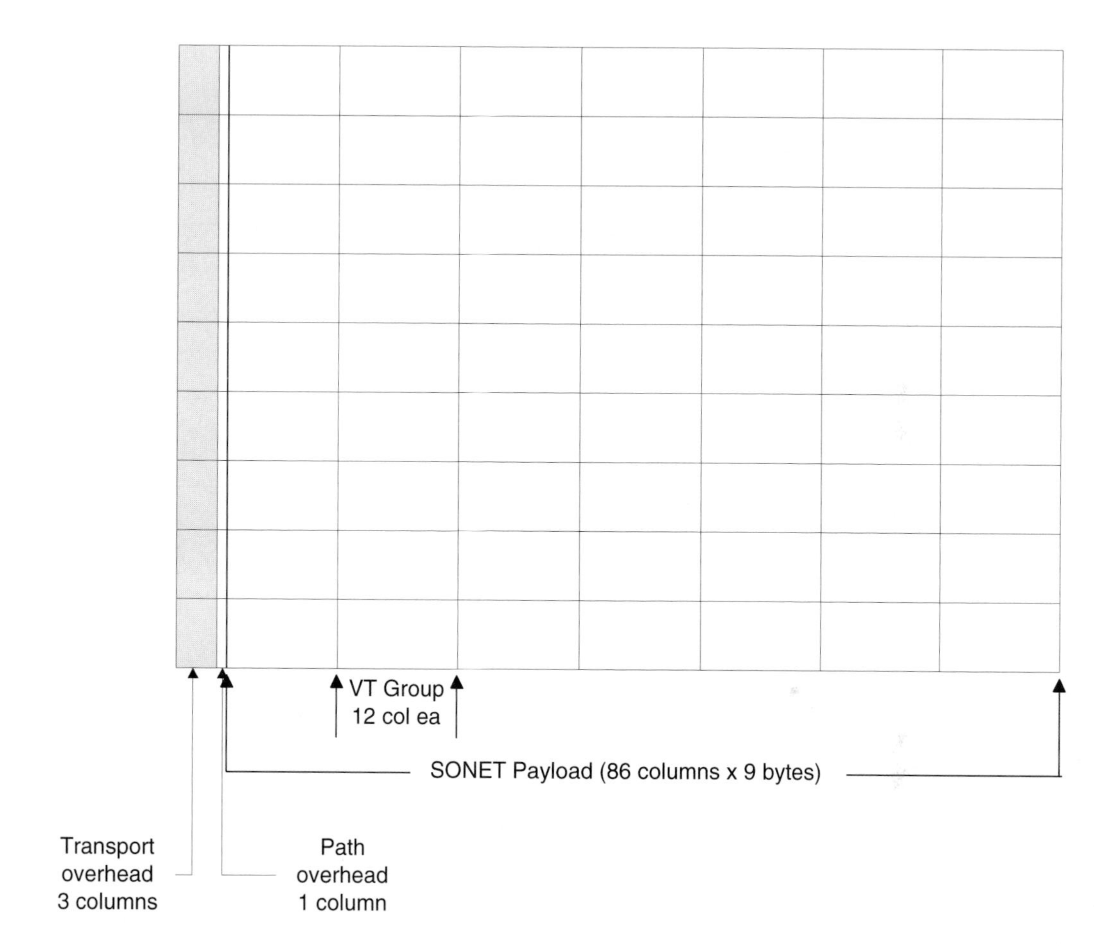

Figure C1.1 SONET frame

use data buffers to handle the unavoidable slips, since the aim was for extremely high data rates (recall that an STS-1 frame would require a 125 μsec buffer). SONET manages offsets in the arrivals of frames without significant buffer delays by launching frames according to the local high-grade clock and allowing an incoming data frame to spill over into a second frame, if necessary. A *pointer* in the SONET header (the "H" bytes in Figure C1.2) lets the system know where the first data byte is in the transport frame. This allows the SPE to be aligned arbitrarily within the transport frame. Offsets between the clock frequencies of different systems are handled by adding a byte whenever required and incrementing the pointer accordingly.

90 BYTES

SPE -- 87 BYTES

Framing A1	Framing A2	STS-1 ID C1	Trace J1	
BIP-8 B1	Order wire E1	USER F1	BIP-8 B3	
Data Comm D1	Data Comm D2	Data Comm D3	Signal Label C2	
Pointer H1	Pointer H2	Pointer Action H3	Path Status G1	
BIP-8 B2	APS K1	APS K2	User Channel F2	
Data Comm D4	Data Comm D5	Data Comm D6	Multi-Frame H4	
Data Comm D7	Data Comm D8	Data Comm D9	Growth Z3	
Data Comm D10	Data Comm D11	Data Comm D12	Growth Z4	
Growth Z1	Growth Z2	Order wire E2	Growth Z5	

SECTION OVERHEAD

LINE OVERHEAD

9 BYTES

TRANSPORT OVERHEAD

PATH OVERHEAD

STS-1 FRAME -- 125 MICROSECONDS

Figure C1.2 SONET header

C2 – Packet Switching, ATM, and Gigabit Ethernet

The synchronous transport systems just described have caused a major improvement in the operation of long distance trunks and high-density interconnects. These systems operate at the highest levels of the communications network, where traffic flow is relatively well defined. *Packet switching* systems

Table C1-2 Common non-synchronous telephony multiplexes

	North America			Europe, Asia, South America		
	Name	**Rate (bps)**	**Multiple**	**Name**	**Rate (bps)**	**Multiple**
Voice rate	DS0	64 k		E0	64 k	
Increasing Multi-Plexes	DS1	1.544 M	24 DS0	E1	2.048 M	32 E0
	DS2	6.312 M	96 DS0	E2	8.448 M	128 E0
	DS3	44.736 M	28 DS1	E3	34.368 M	16 E1

have generated a similar change in the switching of lower rate streams at the interfaces between the high-density backbones and end-users. At these levels, the traffic flow requirements are highly variable. At one extreme are voice, audio, and video, which are very time-sensitive, so they appear to require a locked-up end-to-end circuit to maintain continuous communication. At the other extreme is bursty data that would make very inefficient use of a locked-up circuit.[6] Packet switched networks put traffic into *packets* or *cells*, which can be either fixed length (as in *Asynchronous Transfer Mode* [ATM]) or variable length (as in *Internet Protocol* [IP] *or Frame Relay*). These networks were established specifically to maximize the use of individual communication channels while accommodating a variety of traffic classes. This means that there does not need to be a "video switch" that is different from a "voice switch," etc.

In ATM the individual data streams get a pre-established circuit routing through a call setup process. The route is identified with a number, and each cell carries that ID. When the cell gets to a *router*, the traffic-directing device at a network node point, the ID is read and the routing is retrieved from a locally stored look-up table. Because all ATM cells travel the same route, they are guar-

6. The differences between these two extremes are apparent when one considers the problems that the local telephone companies have had in dealing with the onslaught of Internet browsing. Their system is based on locked-up connections, with capacity sized appropriately for seven-minute-average (voice) phone calls. The unexpected and rapid shift by their customers to using the phone lines for computer modem hook-ups, which can hold the circuit for hours at a time, has caused severe capacity shortages.

anteed to arrive in the order in which they were launched. In IP there is no specific path set up and, in a sense, each packet finds its own way through the network. The packet header carries destination information, and each router decides which output direction is the best—at that moment—to get to the packet's ultimate destination, based on the router's general routing tables. (An interesting comparison between ATM, whose heritage is from international telephony organizations, and IP, which arose with the data-centric Internet, is contained in S.G. Steinberg, "Netheads vs. Bellheads.")[7]

ATM is based on 53-byte cells. One immediately wants to ask why a system based on such short cells (a) doesn't bog down due to switching and (b) doesn't lose throughput efficiency because of the relatively large number of headers. The switching technology, initially called "fast packet switching," was specifically developed not to impede cell movement. The switch inspects the header only long enough to see if this is its destination and, if not, to send it on in the proper direction. The ATM cell header uses five bytes, or less than 10 percent of the full 53-byte cell. The header includes information about cell priority, such as time sensitivity, so an ATM network can offer different *quality of service* (QoS) levels, with appropriately differentiated tariffs.

ATM technology maps into the Data Link Layer (Layer 2) in the OSI seven-layer model (Table C2-1).[8] Before user data can enter an ATM network, it is passed through an *ATM Adaptation Layer* (AAL), which is generally part of the user equipment. The AAL segments the data into 48-byte cell payloads and generates some of the information, including QoS, that will go into the ATM header. The AAL passes its packets on to the *ATM layer*, which generates ATM cell headers and multiplexes the cells into the physical transport layer. The process is reversed at the destination point.

In all packet-based systems, the cells pass through many network nodes between source and destination. In ATM each cell header is given a *virtual path*

7. S.G. Steinberg, "Netheads vs Bellheads" Wired Magazine, October 1996.

8. J.D. Day and H. Zimmerman, "The OSI Reference Model," *Proc. of the IEEE*, vol 71, 1334–40, December 1983

Table C2-1 Open Systems Interconnection Model

OSI Layer	Layer Functions
Application	E-mail, file transfer, terminal emulation
Presentation	Data formatting, compression, encryption
Session	Session set-up and maintenance, handshakes
Transport	End-to-end delivery, packet creation
Network	Routing
Data Link	Framing, error checking
Physical	Actual transport of bits

identifier and a *virtual channel identifier* (VPI/VCI), per the set-up. The VPI identifies the route, and the VCI indicates which other cells may be linked with this one. At each node the cell passes through the ATM layer so that the VPI/VCI information in the header can be read and then translated appropriately for the next node. This is done with a look-up table in the ATM switch. At each port the QoS information is checked and buffering is provided to prevent misordering.

Gigabit Ethernet (1000Base–X) is an alternate packet communications system with a 1 Gbps backbone, which builds directly on the huge installed base of 10Base-T (10 Mbps) Ethernet networks. Broadcast (point-to-multipoint) applications are transported more naturally in Ethernet than in ATM. On the other hand, QoS classification and billing are inherent in ATM, but are still in the process of being adapted to 1000Base-X and other packet systems.

At the present time, ATM and 1000Base-X are being implemented for the WANs that link routers and servers in headends with the external Internet network. At some time in the future it is possible that these high-level transports will reach all the way to end-users via the HFC network. Bear in mind that any communications that pass through routers will find the LAN-like HFC network much more "friendly" than a telephony-style switched network.

C3 – Internet Communications

Nearly everything discussed in this book relates to the *physical layer* (PHY) of a network, the first level of the seven-layer OSI communication system (Table C2-1). The Internet is intended to operate over any physical network, hence no particular PHY specifications are associated with it. This is one of the aspects that has made the Internet ubiquitous. The protocols that have been established for the Internet can be mapped into the OSI model, as shown in Table C3-1.

Table C3-1 Mapping of Internet protocols to the OSI model

OSI LAYER						
	Application	World Wide Web (HTTP)	File Transfer (FTP)	E-Mail (SMTP)	Network Managemt (SNMP)	Addressing (DNS)
	Transport	Transmission Control Protocol (TCP)			User Datagram Protocol (UDP)	
	Network	Internet Protocol (IP)				
	Data link	Ethernet	Token Ring	FDDI	ATM	Other

Internet protocol (IP)

Another feature of the Internet is its combination of the simplicity at its core coupled with the great flexibility afforded to users for making specific extensions, as needed. An example is the Network Layer *Internet Protocol* (IP). It is merely a "best efforts" packet delivery system. Delivery is not guaranteed, nor is the sequencing of a series of packets. What is guaranteed is that if a packet is not delivered to its destination within the prescribed amount of time,[9] it will be deleted from the system: there is no "dead letter office." This means that the system doesn't have to concern itself with packets that could float around endlessly because of an error in addressing or routing.

9. In this case, "not within the prescribed amount of time" really means that it has transited too many router-to-router links. There is a "time-to-live" field in the packet header that gets reduced by one at each router. When the field goes to zero, the router turns the packet back into recyclable electrons.

IP allows variable-length packets (up to 2^{16} bytes, but limited to 576 bytes for the Internet), which means that the header overhead can be low—providing that the packet doesn't get lost.

Transport protocols (TCP and UDP)

The transport layer creates packets including header information, inserts the packets into the system, and extracts them. It also can provide oversight of the delivery, if desired. Two transport protocols are used in the Internet: *Transmission Control Protocol* (TCP) and *User Datagram Protocol* (UDP).

TCP is the protocol used in most of the applications commonly associated with the Internet, such as Web browsing, e-mail, and file transfer. It sets up connections through the network and maintains sequences of data. If a packet does not reach its destination within the allotted time, TCP informs the sender that the transmission failed. UDP, on the other hand, is much more rudimentary. In a sense, UDP pins a destination address onto a packet, puts the packet onto the network, and waves good-bye.

Why would anyone use UDP when TCP offers so much more capability? Most of the UDP users want a robust "bare bones" protocol so that they can provide monitoring and control of the transmission—by their own methods—to the maximum extent. An example is network management. If the information and commands for managing the network are flowing over the network itself, a great deal of care is needed to ensure that all of the data is accurately and assuredly delivered. Thus SNMP programmers need to include error checking (not just for the data, but also for its sequence) and establish a means for generating re-transmission requests when an error is detected.

How a packet gets from source to destination

It is informative to follow a packet from source to destination over the Internet. Viewing the Internet as the simplified network shown in Figure C3.1, we will outline seven of the steps taken in transporting information. This treatment follows closely the presentation by M. H. Ammar of Georgia Tech at the 1997 SCTE Emerging Technologies Conference.[10]

10. M.H. Ammar, "What is TCP/IP?," 1997 Conference on Emerging Technologies, Nashville, TN, January 1997, Society of Cable Telecommunications Engineers, Exton, PA.

1. The application instructs TCP to deliver data to the destination address.
2. TCP constructs a packet and instructs IP to deliver it to the destination address.
3. IP instructs Ethernet to deliver the packet to the Ethernet interface at the router.
4. Network 1 delivers the packet to a router's Ethernet interface.
5. Ethernet delivers the packet to IP.
6. IP makes a routing decision, based on its routing table.
7. Process continues across network 2.

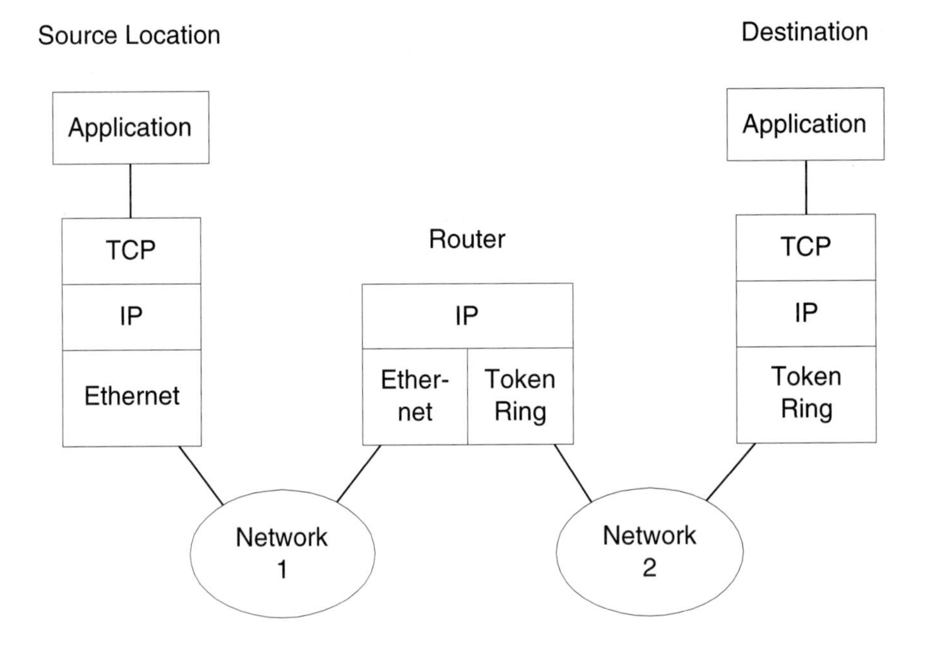

Figure C3.1 Internet network stack

Note that the network interfaces in the second network are token ring, rather than Ethernet, which is an example of the flexibility of the Internet.

In an actual transaction, the packet will travel through many routers. It is obvious that the Internet puts large demands on router capacity and speed. Accordingly, the development of router technology is moving very rapidly.

APPENDIX D

Useful Conversions and Tables

Decibels

Cable engineers constantly deal with devices such as cable, fiber, and taps that attenuate signals by some fixed fraction. Since logarithmic addition and subtraction are a lot more handy than multiplication and division, the industry long ago adopted—from radio engineers—the use of decibels to describe signal strengths (dBmV or dBμV), powers (dBm), and attenuations (dB).

Voltage

The basic definition for signal strength in dB millivolts (dBmV) is

$$\text{E expressed in dBmV} = 20*\log[\text{V expressed in mV}].$$

Examples:

1 volt = 1000 mV = 60 dBmV

1 mV = 0 dBmV

30 dBmV = $10^{(30/20)}$ mV = 31.6 mV

In many countries of Europe and other parts of the world, it is common to express signal strengths in dB microvolts (dBμV). The correspondence is

$$\begin{aligned}\text{E in dB}\mu\text{V} &= 20 * \log(\text{V in }\mu\text{V}) = 20 * \log(1000 * \text{V in mV}) \\ &= 20 * \log(\text{V in mV}) + 60 = \text{E in dBmV} + 60\end{aligned}$$

As mentioned, decibels make it easy to multiply two quantities. On the other hand, adding or subtracting two signal levels expressed as decibels generally requires a look-up table, such as Table D1.

Table D-1 Adding two voltage ratios, such as signal levels. Find the box corresponding to the difference between the two signal levels and add the amount in the table to the larger of the two individual levels.

Δ	0.0	0.1	0.2	0.3	0.4	0.5	0.6	0.7	0.8	0.9
0.0	6.02	5.97	5.92	5.87	5.82	5.77	5.73	5.68	5.63	5.58
1.0	5.53	5.49	5.44	5.39	5.35	5.30	5.26	5.21	5.17	5.12
2.0	5.08	5.03	4.99	4.95	4.90	4.86	4.82	4.78	4.73	4.69
3.0	4.65	4.61	4.57	4.53	4.49	4.45	4.41	4.37	4.33	4.29
4.0	4.25	4.21	4.17	4.13	4.10	4.06	4.02	3.98	3.95	3.91
5.0	3.88	3.84	3.80	3.77	3.73	3.70	3.66	3.63	3.60	3.56
6.0	3.53	3.50	3.46	3.43	3.40	3.36	3.33	3.30	3.27	3.24
7.0	3.21	3.18	3.15	3.12	3.09	3.06	3.03	3.00	2.97	2.94
8.0	2.91	2.88	2.85	2.83	2.80	2.77	2.74	2.72	2.69	2.66
9.0	2.64	2.61	2.59	2.56	2.53	2.51	2.48	2.46	2.44	2.41
10.0	2.39	2.36	2.34	2.32	2.29	2.27	2.25	2.22	2.20	2.18
11.0	2.16	2.13	2.11	2.09	2.07	2.05	2.03	2.01	1.99	1.97
12.0	1.95	1.93	1.91	1.89	1.87	1.85	1.83	1.81	1.79	1.77
13.0	1.75	1.74	1.72	1.70	1.68	1.67	1.65	1.63	1.61	1.60
14.0	1.58	1.56	1.55	1.53	1.51	1.50	1.48	1.47	1.45	1.44
15.0	1.42	1.41	1.39	1.38	1.36	1.35	1.33	1.32	1.31	1.29
16.0	1.28	1.26	1.25	1.24	1.22	1.21	1.20	1.19	1.17	1.16
17.0	1.15	1.14	1.12	1.11	1.10	1.09	1.08	1.06	1.05	1.04
18.0	1.03	1.02	1.01	1.00	0.99	0.98	0.96	0.95	0.94	0.93
19.0	0.92	0.91	0.90	0.89	0.88	0.87	0.86	0.86	0.85	0.84
20.0	0.83	0.82	0.81	0.80	0.79	0.78	0.77	0.77	0.76	0.75

Example: To combine signals of 45.2 and 42.1 dBmV. The difference between the two is 3.1. Find 4.61 at the intersection of the 3.0 row and the 0.1 column. The sum of the two signals is 45.2 + 4.6 = 49.8 dBmV.

Table D-1 can also be used for combining any other quantities that add on a 20log basis, such as CTB and crossmodulation, but care must be taken with the negative signs (both CTB and XMOD are expressed as negative quantities). The best way to keep this straight is to remember that combining two CTB values, for example, will make the worse of the two numbers even worse. Thus Table D-1 is used as follows:

1. Take the absolute difference of the two CTBs (ignoring minus signs).
2. Find the correction factor for that difference in Table D-1.
3. Add that number to the worse CTB (the negative number with the smaller absolute value).

Example: Combine CTBs of –65.2 and –62.1.

1. The absolute difference is (65.2 – 62.1) = 3.1
2. From Table D-1 the correction factor is 4.6 (at the intersection of the 3.0 row and the 0.1 column).
3. The combined CTB is the - 62.1 + 4.6 = - 57.5.

Power

Power ratios are defined in decibels as

$$\text{Power ratio} = 10*\log [P_2/P_1].$$

Examples:

Ideal power splitter: $10*\log[\text{Power}_{out}/\text{Power}_{in}] = 10*\log[0.5/1] = -3$ dB

Practical power splitter = ideal splitter less approximately 0.5 additional loss = –3.5 dB

Power in dB milliwatts (dBm) is defined as

$$\text{Power in dBm} = 10*\log[P \text{ (in mW)}].$$

Throughout this text we have referred to power in terms of dBmV. Since dBmVs are defined relative to a voltage, one could ask how it is possible to use voltage to describe powers. The reason is that everything in the cable engineering environment is always referenced to a 75 ohm impedance, so there is a one-to-one correspondence between voltage and power. This can be show mathematically by the following:

If the impedance of a system is Z ohms (Ω), then the power through the system is V^2/Z, where V is in volts. Therefore:

$$\text{Power in mW} = 1000*\text{Power in W} = 1000*V^2/Z$$
$$\text{Power in dBm} = 10*\log(\text{Power in mW})$$
$$= 10*\log(1000*V^2/Z) = 30 + 20*\log(V) - 10*\log(Z)$$
$$= 30 + 20*\log[(V \text{ in mV})/1000] - 10*\log(Z)$$
$$= 30 + 20*\log[(V \text{ in mV})] - 60 - 10*\log(Z)$$
$$= \text{dBmV} - 30 - 10*\log(Z)$$

Since the impedance throughout a cable system is always 75 Ω, this becomes

Power in dBm = dBmV - 48.75 **(75 Ω).**

As long as the impedance is 75 Ω, then there is this simple correspondence between voltage and power.

Examples:
48.75 dBmV = 0 dBm
30 dBmV = -18.75 dBm
-30 dBm = 18.75 dBmV

Some laboratory instruments are calibrated for 50 Ω. For such instruments the conversion is

Power in dBm = dBmV - 46.99 **(50 Ω).**

Adding powers expressed in decibels is aided by Table D-2.

Table D-2 can also be used for combining quantities that add on a 10*log basis, such as C/N and CSO. As was the case with CTB and XMOD, however, one needs to be careful with signs.[1] Again, the best way to do this is to remember that—in the case of C/N—you are adding noise, which means that the aggre-

1. In the case of C/N we are actually adding inverses:

$$\text{Combined C/N} = \frac{\text{Carrier}}{\text{Total noise}} = \text{Carrier}/(\text{Noise}_1 + \text{Noise}_2)$$

$$\frac{1}{\text{C/N}} = \frac{\text{Noise}_1 + \text{Noise}_2}{\text{Carrier}} = \frac{N_1}{C} + \frac{N_2}{C} = \frac{1}{(\text{C/N})_1} + \frac{1}{(\text{C/N})_2}$$

Table D-2 The sum of two power ratios. Find the box corresponding to the difference between the two powers and add the amount in the table to the larger of the two.

Δ	0.0	0.1	0.2	0.3	0.4	0.5	0.6	0.7	0.8	0.9
0.0	3.01	2.96	2.91	2.86	2.81	2.77	2.72	2.67	2.63	2.58
1.0	2.54	2.50	2.45	2.41	2.37	2.32	2.28	2.24	2.20	2.16
2.0	2.12	2.09	2.05	2.01	1.97	1.94	1.90	1.87	1.83	1.80
3.0	1.76	1.73	1.70	1.67	1.63	1.60	1.57	1.54	1.51	1.48
4.0	1.46	1.43	1.40	1.37	1.35	1.32	1.29	1.27	1.24	1.22
5.0	1.19	1.17	1.15	1.12	1.10	1.08	1.06	1.04	1.01	0.99
6.0	0.97	0.95	0.93	0.91	0.90	0.88	0.86	0.84	0.82	0.81
7.0	0.79	0.77	0.76	0.74	0.73	0.71	0.70	0.68	0.67	0.65
8.0	0.64	0.63	0.61	0.60	0.59	0.57	0.56	0.55	0.54	0.53
9.0	0.51	0.50	0.49	0.48	0.47	0.46	0.45	0.44	0.43	0.42
10.0	0.41	0.40	0.40	0.39	0.38	0.37	0.36	0.35	0.35	0.34
11.0	0.33	0.32	0.32	0.31	0.30	0.30	0.29	0.28	0.28	0.27
12.0	0.27	0.26	0.25	0.25	0.24	0.24	0.23	0.23	0.22	0.22
13.0	0.21	0.21	0.20	0.20	0.19	0.19	0.19	0.18	0.18	0.17
14.0	0.17	0.17	0.16	0.16	0.15	0.15	0.15	0.14	0.14	0.14
15.0	0.14	0.13	0.13	0.13	0.12	0.12	0.12	0.12	0.11	0.11
16.0	0.11	0.11	0.10	0.10	0.10	0.10	0.09	0.09	0.09	0.09
17.0	0.09	0.08	0.08	0.08	0.08	0.08	0.07	0.07	0.07	0.07
18.0	0.07	0.07	0.07	0.06	0.06	0.06	0.06	0.06	0.06	0.06
19.0	0.05	0.05	0.05	0.05	0.05	0.05	0.05	0.05	0.05	0.04
20.0	0.04	0.04	0.04	0.04	0.04	0.04	0.04	0.04	0.04	0.04

Example: To add 49.6 and 48.1. The difference between the two is 1.5. Find 2.32 at the intersection of the 1.0 row and the 0.5 column. The sum of the two powers is 49.6 + 2.3 = 51.9.

gate system C/N must be worse than the worse of the two C/Ns you are combining.

Example: Combining C/Ns of 49.6 and 48.1
The difference is 1.5

Find 2.32 in Table D-2 (at the intersection of the 1.0 row and the 0.5 column)
The worse C/N is 48.1
The combined C/N is 48.1 – 1.5 = 46.6.

Reflections, Return Loss, and Flatness

When an electromagnetic wave reflects off a discontinuity, it will cause an interference with the incoming wave. This will lead to peaks and valleys, which the cable TV industry quantifies in a *flatness* measurement that is generally expressed either as X dB peak-to-valley or as ±X/2 dB. The industry also uses the term *return loss* to refer to the number of dB by which the power in the reflected signal is lower than the incoming signal:

$$\text{Return loss} = 10*\log[P_{incoming} / P_{reflected}]$$

$$= 20*\log[E_{in} / E_{refl}]$$

Note that return loss is a positive number[2] since the reflected signal will always be lower than the incoming one.

Let's calculate the peak-to-valley for a typical return path case: a 125-ft drop cable between a tap with return loss of RL_{tap} and an in-home combining device with return loss of RL_{home}. The upstream signals from the home will be interfered with by signals that emerged from the home earlier and that were reflected backward by the tap and then back upstream by the home device. For simplicity we will assume that the signals transmitted from the home are at all return frequencies and at a uniform signal level E_0. A round trip of an RF signal down and up a 125 ft (37 m) drop cable takes approximately 0.28 µsec, which is equivalent to a full period at 3.6 MHz. Thus at the tap there will be a maximum signal at frequencies equal to integer multiples of 3.6 MHz, due to the coherent adding, with minimum signals halfway in-between, due to subtracting.

We will use the symbol CL_{40} for the cable loss at 40 MHz and ignore frequency differences in cable loss, since these differences are small over the return spectrum. The maximum and minimum signals at the tap will then be

2. While there is not universal agreement on this signing convention, it does appear logical that the <u>loss</u> should be positive.

$$\begin{aligned} E_{net} &= (E_0 - CL_{40}) \pm (E_0 - 3{*}CL_{40} - RL_{tap} - RL_{home}) \\ &= (E_0 - CL_{40}) \times [1 \pm (2{*}CL_{40} + RL_{tap} + RL_{home})]. \end{aligned}$$

The question is thus reduced to adding two voltages that are expressed in dB. We will use Table D-1 for the addition. The peak difference Δ between the original signal and the level formed by the interference of the two signals is the term in the inner round parentheses, $2{*}CL_{40} + RL_{tap} + RL_{home}$.

Let's assume some values. At 40 MHz the loss of drop cable is about 0.008 dB/ft, so a 125 ft length will have 1.0 dB of loss. If there is a forward-band trapping device on the tap, the return loss in the return band could be as low as 5 dB. The home device should have a minimum return loss of 10 dB. Combining these values gives Δ= 2*1 + 5 + 10 = 17 dB. From Table D-1, we can see that the peak will be 1.15 dB above $(E_0 - CL_{40})$. This means that the peak-to-valley is twice this, or 2.3 dB.

For those readers coming from industries that use other terminology, we need to provide the appropriate conversion formulas.

The term *voltage standing wave ratio* (VSWR, pronounced "viz-wahr") is used by many electrical engineers to describe the interference between an incoming wave and a reflected wave. For a continuous pure tone, it is easy to see that there will be destructive interference (causing a minimum amplitude E_{min}) at some distance from the reflecting interface. At another point there will be constructive interference, producing a maximum signal level E_{max}. VSWR is the ratio of the peak signal to the valley signal:

$$VSWR = E_{max}/E_{min}$$

VSWR is a positive number, greater than 1.

The *reflection coefficient* is the excursion (half the peak-to-valley) divided by the average, sometimes expressed as a percentage:

$$R = \frac{\text{Excursion}}{\text{Average}} = \frac{(E_{max} - E_{min})/2}{(E_{max} + E_{min})/2} = \frac{E_{max} - E_{min}}{E_{max} + E_{min}}$$

Percent reflection is defined as 100 × R

Let's convert a10 dB return loss in our example above into its equivalent expressed as a VSWR and as a reflection coefficient:

$$E_{refl} = 10^{-RL/20} \times E_{in} = 10^{-0.5} \times E_{in}$$

$$E_{max} = E_{in} + E_{refl} = (1 + 10^{-0.5}) \times E_{in}$$

$$E_{min} = E_{in} - E_{refl} = (1 - 10^{-0.5}) \times E_{in}$$

$$VSWR = \frac{E_{max}}{E_{min}} = \frac{(1 + 10^{-0.5}) \times E_{in}}{(1 - 10^{-0.5}) \times E_{in}} = \frac{1.32}{0.68} = 1.9$$

$$R = \frac{(E_{max} - E_{min})}{(E_{max} + E_{min})} = \frac{(1 + 10^{-0.5}) \times E_{in} - (1 - 10^{-0.5}) \times E_{in}}{(1 + 10^{-0.5}) \times E_{in} + (1 - 10^{-0.5}) \times E_{in}}$$

$$= \frac{1.32 - 0.68}{1.32 + 0.68} = \frac{0.64}{2} = 0.32$$

The conversion formulas for these quantities are given in Table D-3.P

Table D-3 Conversions between reflection measurements

	Expressed in terms of:			
Quantity:	Return Loss	VSWR	R	%R
Return Loss		$20*\log\left(\frac{(VSWR+1)}{(VSWR-1)}\right)$	$-20*\log R$	$-20*\log(\%R/100)$
VSWR	$\frac{1 + 10^{-RL/20}}{1 - 10^{-RL/20}}$		$\frac{(1 + R)}{(1 - R)}$	$\frac{(1 + \%R/100)}{(1 - \%R/100)}$
R	$10^{-RL/20}$	$\frac{(VSWR-1)}{(VSWR+1)}$		$\%R/100$
%R	$100*10^{-RL/20}$	$\frac{100(VSWR-1)}{(VSWR+1)}$	$100*R$	

List of Acronyms

Acronym	Meaning	Chapter	Type
%R	Percent reflection	D	Cable system parameter
AAL	ATM adaptation layer	C	ATM term
AC	Alternating current	6	Electric power term
ADM	Add/drop multiplexer	12, C	Digital transport subsystem
AGC	Automatic gain control	2, 4, 11	Amplifier subsystem
AM	Amplitude modulation	2	Analog modulation technique
AML	AM microwave link	2	Transport method
AM-VSB	AM vestigial sideband	2	Composite video modulation technique
ASK	Amplitude shift keying	3	Digital modulation method
ATM	Asynchronous transfer mode	C	Packet communications protocol
AWGN	Additive white gaussian noise	3	Communication impairment
BER	Bit error rate	3	Digital performance parameter
BML	Business management layer	13	NMS term
BPF	Bandpass filter	3	Passive device
BPS	Bits per second	3, 8	Communication unit
BPSK	Binary phase shift keying	3	Digital modulation method
BTS	Base transceiver substation	8	PCS subsystem
BW	Bandwidth	4, 3	Communication system parameter
C/N	Carrier to noise ratio (in dB)	4, 3	System performance parameter
CAD	Computer-aided design	1	Electronic design platform
CATV	Community antenna television	1	Original term for cable TV
CB	Citizens band	4	Portion of RF spectrum
CCS	Hundred call seconds	8	Traffic measurement unit
CDMA	Code division multiple access	8	Multiple access technique
CIN	Composite intermodulation noise	4	Distortion parameter

Acronym	Meaning	Chapter	Type
CM	Cable modem	1, 8, 11	Consumer data device
CMI	Cable microcell integrator	8	PCS subsystem
CMTS	Cable modem termination system	8, 12	Cable modem headend subsystem
CO	Central office	8	Telephone headend
CSO	Composite second order	4, 10, D	Distortion parameter
CTB	Composite triple beat	4, 10, D	Distortion parameter
CW	Continuous wave	10	Unmodulated sinewave
DA	Distribution amplifier	2, 4	RF network component
DAVIC	Digital Audio-Visual Council	8	Standards organization
dB	Decibel	D	Logarithmic expression of ratio
dBc	Decibel relative to carrier	5	Logarithmic expression of ratio
dBm	Decibel relative to one milliwatt	D	Logarithmic expression of power ratio
dBmV	Decibel relative to one millivolt	D, 4	Logarithmic expression of voltage ratio
dBμV	Decibel relative to one microvolt	D, 4	Logarithmic expression of voltage ratio
DBS	Direct broadcast satellite	1	Microwave delivery system
DC	Direct current	2, 6	Electric power term
DFB	Distributed feedback	9	Laser type
DNS	Domain name system	C	Internet subsystem
DOCSIS	Data Over Cable Service Interface Specification	8	Cable modem system specification
E_b/N_o	E-b over N-zero	3	Communication system parameter
EDFA	Erbium-doped fiber amplifier	12	1550 nm optical amplifier
EM	Element manager	13	NMS term
EQ	Equalizer	5, 3, 11	Device
FCC	Federal Communications Commission	4	U.S. government organization
FDM	Frequency division multiplex	2	Method for transporting multiple signals
FDMA	Frequency division multiple access	8	Multiple access technique
FEC	Forward error correction	3	Digital transport enhancement

Acronym	Meaning	Chapter	Type
FM	Frequency modulation	2	Analog modulation technique
FP	Fabry-Perot	9	Laser type
FSK	Frequency shift keying	3	Digital modulation method
GPS	Global Positioning System	C	Satellite reference system
HDT	Host digital terminal	8	Telephony subsystem
HF	High frequency	4	Portion of RF spectrum
HFC	Hybrid fiber/coax	2	Cable distribution type
HIC	Headend interface converter	8	PCS component
HP	Homes passed	12, 8	Cable system parameter
HPN	Homes per node	8	System parameter
HTTP	Hypertext transfer protocol	C	Internet protocol
IC	Integrated circuit	4, 7	Semiconductor circuit
ICS	Ingress control switch	13, 11	Amplifier component
IEEE	Institute of Electrical and Electronics Engineers	8	International engineering organization
IF	Intermediate frequency	2	RF frequency
IIN	Interferomentric intensity noise	9	Fiberoptic system impairment
INMS	Integrated network management system	13	High-level NMS
IP	Internet protocol	C, 8	Packet data protocol
IPG	Interactive program guide	1	Service offering
IPPV	Impulse pay per view	1	Service offering
IRD	Integrated receiver/decoder	2	Headend component
IRT	Integrated receiver/transcoder	2	Headend component
ISP	Internet service provider	12	Access operator
ITU	International Telecommunication Union	8, C	Standards organization
kbps	Kilobits per second	3	Data transmission rate
kHz	Kilohertz	4, A	Frequency unit
km	Kilometer	2	Length unit

Acronym	Meaning	Chapter	Type
kWh	Kilowatt-hour	6	Energy unit
LAN	Local area network	8, 13	Network type
LE	Line extender	2	RF network component
Mbps	Megabits per second	3, 8, C	Data transmission rate
MCNS	Multimedia Cable Network System	8	Cable standards consortium
MHz	Megahertz	2	Frequency unit
MIB	Management information base	13	SNMP term
MMDS	Multichannel Multipoint Distribution Service	1	Microwave delivery system
MPN	Mode partition noise	9	Fiberoptic system impairment
M-QAM	M-ary quadrature amplitude modulation	3	Digital modulation method
MSC	Mobile swiching center	8	PCS subsystem
MSO	Multiple system operator	12	Cable communications corporation
MTBF	Mean time between failures	7	Reliability parameter
MTTR	Mean time to repair	7	Reliability parameter
MVP	Modulating video processor	2	Scrambling system component
NCTA	National Cable Television Association	Preface	North American cable organization
NE	Network element	13	NMS term
NF	Noise figure	4	RF performance parameter
NGDC	National Geophysical Data Center	4	U.S. government organization
NIU	Network interface unit	8, 5	Side-of-house box
NMS	Network management system	13	Generic term
NPR	Noise power ratio	4, 10	RF performance parameter
NTSC	National Television Systems Committee	2, 4	Broadcast TV standard
OC-n	Optical carrier, nth level	C	Element in SONET and SDH hierarchy
OFDM	Orthogonal frequency division multiplexing	8	Multiple access technique
OMI	Optical modulation index	9	Laser parameter
OOK	On-off keying	3	Digital modulation method

Acronym	Meaning	Chapter	Type
OSI	Open Systems Interconnection	C	Standardized structure for network architectures
PAL	Phase-alternating line	4	Broadcast TV standard
PC	Personal computer	1, 8	End-user device
PCS	Personal communications services	1, 8	Wireless communication system
PDF	Probability density function	10	Statistical parameter
PEP	Peak envelope power	4	Radiation unit
PHY	Physical layer	C	Element of OSI hierarchy
PS	Power Supply	6, 7	Distribution plant element
PSK	Phase shift keying	3	Digital modulation method
PSTN	Public switched telephone network	8, 12	Telephone industry term
QAM	Quadrature amplitude modulation	3	Digital modulation method
QoS	Quality of service	C	Service parameter
QPSK	Quadrature phase shift keying	3	Digital modulation method
R	Reflection coefficient	D	Cable system parameter
RAD	Remote antenna driver	8	Early PCS term
RASP	Remote antenna signal processor	8	Early PCS term
RF	Radio frequency	2	Portion of electromagnetic spectrum
RIN	Relative intensity noise	9	Laser parameter
RMS	Root mean square	10	Averaging method
ROM	Read-only memory	3	Computer subsystem
RPD	Return path demodulator	11, 12	Headend component
RSA	Return step attenuator	5	Passive device
RT	Remote terminal	8	Telephone network subsystem
SCTE	Society of Cable Telecommunications Engineers	Acknowl-edgments	North American cable organization
SDH	Synchronous digital hierarchy	C	Digital transport standard
SML	Service management layer	13	NMS term
SMTP	Simple Mail Transport Protocol	C	E-mail protocol

Acronym	Meaning	Chapter	Type
SNMP	Simple Network Management Protocol	C	NMS term
SNR	Signal to noise ratio (in dB)	3	Communication system parameter
SONET	Synchronous optical network	C, 12	Digital transport standard
SPE	Synchronous payload envelope	C	SONET term
STB	Set-top box	1, 11	Terminal equipment
STM-n	Element in SDH hierarchy	C	Digital transport frame
STS-m	Element in SONET hierarchy	C	Digital transport frame
TCI	Tele-Communications, Inc.	10	US MSO
TCP	Transmission control protocol	C	Internet protocol
TDM	Time division multiplex	3, 8	Method for transporting multiple signals
TDMA	Time division multiple access	8	Multiple access technique
TEC	Thermoelectric cooler	9	Laser subsystem
TMN	Telecommunications Management Network	13	NMS hierarchy
TP	Test point	11, B	Amplifier component
TV	Television	2, 4	End-user device
UDP	User datagram protocol	C	Internet protocol
VAC	Volts AC	6	Electrical unit
VCI	Virtual channel identifier	C	ATM term
VPI	Virtual path identifier	C	ATM term
VSWR	Voltage standing wave ratio	D	Reflection parameter
VT	Virtual tributary	C	SONET term
WAN	Wide area network	8, 13	Extensive network
XMOD	Cross modulation	4, D	Distortion parameter

Index

E

F

J

L

M

N

O

P

T

U

V

W

Z